The Art and Ethics of Generative AI

Generative artificial intelligence (AI) technologies are reshaping the way we create, communicate, and compute, presenting both unprecedented opportunities and profound challenges. As these systems increasingly influence decisions that shape our lives, it becomes clear that performance metrics alone cannot define their success. A deeper understanding of the intersection between technical innovation and human values is essential to navigate this transformative era.

The Art and Ethics of Generative AI: Fairness, Transparency, and Human Values provides a balanced exploration of generative AI, combining technical insights with ethical considerations. It explains generative model architectures and learning paradigms in an accessible way while addressing critical issues such as algorithmic bias, explainability challenges, and emerging regulatory approaches. Practical frameworks for responsible AI development are offered, alongside multidisciplinary perspectives that bridge technical and humanistic viewpoints. Abstract concepts are brought to life through practical illustrations, demonstrating how ethical considerations play out in real-world AI applications.

Designed for a wide audience, the book serves AI practitioners seeking actionable guidance for building responsible systems, researchers exploring nuanced ethical dimensions, educators and students looking for comprehensive learning materials, and policymakers needing governance insights. It is an essential resource for anyone engaged in the rapidly evolving AI ecosystem.

The Art and Ethics of Generative AI

Fairness, Transparency, and Human Values

Kirti Seth, Nidhi Gupta and Ashish Seth

CRC Press
Taylor & Francis Group
Boca Raton London New York

CRC Press is an imprint of the
Taylor & Francis Group, an **informa** business

Designed cover image: Shutterstock

First edition published 2027
by CRC Press
2385 NW Executive Center Drive, Suite 320, Boca Raton FL 33431

and by CRC Press
4 Park Square, Milton Park, Abingdon, Oxon, OX14 4RN

CRC Press is an imprint of Taylor & Francis Group, LLC

ISBN: 978-1-041-19735-5 (hbk)
ISBN: 978-1-041-19884-0 (pbk)
ISBN: 978-1-003-71395-1 (ebk)

DOI: 10.1201/9781003713951

Typeset in Times
by Apex CoVantage, LLC

Contents

Preface xiii
About the Authors xv
List of Abbreviations xvii

Chapter 1 Foundations of Generative AI 1

1.1 AI and Its Evolution 1
1.2 Understanding Generative AI, Fairness, and Explainability 3
1.2.1 Fairness in GenAI 3
1.2.2 Explainability in GenAI 3
1.2.3 The Interconnection of GenAI, Fairness, and Explainability 4
1.3 Historical Development of GenAI 4
1.4 AI and GenAI 7
1.5 AI Pillars: ML, DL, GenAI, and LLM 8
1.5.1 Machine Learning 8
1.5.2 Deep Learning 9
1.5.3 Large Language Models 9
1.6 GenAI Blueprint 9
1.6.1 Input Training Data 10
1.6.2 Foundation Model Training 10
1.6.3 Generation Phase 11
1.7 Core Foundations of GenAI: Mathematics, Models, and Metrics 12
1.7.1 Mathematical Foundations 12
1.7.2 Optimization 14
1.7.3 Foundation Models of ML and DL Relevant to GenAI 14
1.7.4 Autoencoders 16
1.7.5 Recurrent Neural Networks, LSTMs, and GRUs 18
1.8 GenAI Engine: Transformers 19
1.8.1 Attention Mechanism 20
1.9 Loss Functions in GenAI Models 21
1.9.1 Cross-Entropy Loss 21
1.9.2 Kullback–Leibler (KL) Divergence 21
1.10 Metrics for Evaluating Generative Models 21
1.10.1 Perplexity 21
1.10.2 Bilingual Evaluation Understudy (BLEU) 21
1.10.3 Recall-Oriented Understudy for Gisting Evaluation (ROUGE) 22

1.10.4 Fréchet Inception Distance (FID) 22
1.10.5 Inception Score (IS) 22
1.10.6 Human Evaluation 22
1.11 Ethical and Societal Considerations 22
1.11.1 Deepfakes and Misinformation 22
1.11.2 Bias in Generative Models 22
1.11.3 Responsible AI and Alignment 23
Exercises 23
References 25

Chapter 2 The Ethical Landscape of GenAI 27

2.1 The Promise of GenAI 28
2.1.1 Scientific Advancement 29
2.1.2 Creativity and Culture 30
2.1.3 Efficiency, Productivity, and Automation 30
2.1.4 Ethical Challenges and Implications 32
2.2 Bias and Discrimination 33
2.2.1 Misinformation and Deepfakes 33
2.2.2 Intellectual Property and Authorship 33
2.2.3 Privacy and Data Protection 34
2.2.4 Labor Displacement and Economic Inequality 34
2.2.5 Transparency and Accountability 34
2.2.6 Environmental Impact 35
2.2.7 Governance and Global Regulation 35
2.2.8 Ethical Trade-Offs 35
2.3 Pros and Cons of AI Adoption Across Industries 37
2.3.1 Healthcare 37
2.3.2 Finance 37
2.3.3 Education 38
2.3.4 Governance and Law 38
2.3.5 Media and Communication 41
2.3.6 Creative Industries 44
2.3.7 Scientific Research 46
2.3.8 Comparative Synthesis 48
2.4 Ethical Implications Across Industries 49
2.4.1 Healthcare and Biomedicine 49
2.4.2 Finance and Banking 49
2.4.3 Education and Academia 50
2.4.4 Creative Industries and Media 50
2.4.5 Governance, Law, and Security 50
2.4.6 Cross-Sectorial Synthesis 51
2.5 Conclusion 51
Exercises 52
References 53

Chapter 3 Exploring AI Learning Paradigms: From Supervised to Generative Learning....59

3.1 AI Learning Paradigms 59
3.2 Supervised Learning 59
3.3 Unsupervised Learning 61
3.4 Semi-Supervised Learning 63
3.5 Reinforcement Learning 63
3.6 Self-Supervised Learning 68
3.6.1 Algorithms for Self-Supervised Learning 69
3.7 Generative Learning 70
3.8 Summarization and Comparison of Different Learning Paradigms 71
3.9 Emerging and Hybrid Learning Paradigms 73
3.9.1 Transfer Learning and Fine-Tuning 73
3.9.2 Federated Learning 73
3.9.3 Active Learning 73
3.9.4 Meta-Learning (Learning to Learn) 74
3.9.5 Continual and Lifelong Learning 74
3.9.6 Multimodal Learning (Text + Image + Audio) 74
3.9.7 Neuro-Symbolic and Hybrid Models 74
3.10 Case Studies 74
3.10.1 Supervised Learning Case Study—Diagnosis of Disease in Healthcare 74
3.10.2 Unsupervised Learning Case Study—Student Clustering in Education 75
3.10.3 Semi-Supervised Learning Case Study—Medical Image Segmentation 75
3.10.4 Reinforcement Learning Case Study—Autonomous Vehicle Control 76
3.10.5 Case Study on Self-Supervised Learning—Medical Report Generation 76
Exercises 77
References 78

Chapter 4 GenAI: Models and Architecture....81

4.1 GenAI Five-Layer Framework Architecture 81
4.1.1 Data Processing Layer 81
4.1.2 Generative Model Layer 81
4.1.3 Feedback and Continuous Learning Layer 81
4.1.4 Deployment and Integration Layer 82
4.1.5 Monitoring and Maintenance Layer 82
4.2 GenAI Architecture Models 83
4.2.1 Generative Adversarial Networks 83
4.2.2 Variational Autoencoders 84

4.2.3 Transformer-Based Models 86
4.2.4 Diffusion Models 87
4.3 Real-World Applications and Use Cases of GenAI 89
4.3.1 Healthcare 89
4.3.2 Entertainment (Media, Games, Films) 89
4.3.3 Finance 90
4.3.4 Notable Use Cases in Content Creation: Images, Video, Text 90
4.4 Conclusion 91
Exercises 91
References 92

Chapter 5 Fairness in GenAI 93

5.1 Bias in Training Data Description 94
5.1.1 Algorithmic Discrimination 95
5.1.2 Bias in Human Annotations Description 96
5.1.3 Systemic and Feedback Bias 97
5.1.4 Cross-Cultural and Linguistic Bias 97
5.1.5 Intersectional Bias 97
5.2 Approaches to Promote Fairness 98
5.2.1 Algorithmic Debiasing Techniques 98
5.2.2 Promoting Equitable Access 98
5.3 Fairness Measuring Techniques 99
5.4 Case Studies of Bias in GenAI 101
5.4.1 Text Generation Bias: GPT in Hiring Simulations 101
5.4.2 Image Generation Bias: DALL·E and Occupational Stereotypes 101
5.4.3 Chatbot Conversations: Reinforcement of Racial Stereotypes 102
5.4.4 Educational Tools: Gendered Language in Writing Assistants 102
5.4.5 Healthcare Applications: Underrepresentation in Medical Generative Models 103
5.5 Metrics to Evaluate Fairness in GenAI 103
5.5.1 Demographic Parity 103
5.5.2 Equalized Odds 105
5.5.3 Counterfactual Fairness 105
5.5.4 Individual Fairness 106
5.5.5 Representation Parity and Coverage Metrics 106
5.5.6 Embedding and Latent Space Fairness 106
5.5.7 Tools for Fairness Measurement in GenAI 107
5.6 Legal, Ethical, and Policy Dimensions of Fairness in GenAI 107
5.6.1 Legal Frameworks and Regulatory Trends 107
5.6.2 Core Ethical Principles in GenAI Fairness 108

5.6.3 Risks of Non-Compliance and Ethical Negligence 109
5.6.4 Toward a Governance Model for Fair GenAI 109
Exercises 110
References 111

Chapter 6 Opening the Black Box: Explainability in GenAI 113

6.1 Importance of Explainability in GenAI 113
6.1.1 Trust and Transparency 114
6.1.2 Ethical Considerations 114
6.1.3 Regulatory Compliance 115
6.1.4 Debugging and Model Enhancement 115
6.1.5 User Acceptance and Uptake 116
6.2 Challenges in Explaining GenAI 117
6.2.1 Complexity of Deep Neural Architectures 117
6.2.2 Trade-Off between Accuracy and Interpretability 118
6.2.3 Domain-Specific Barriers (e.g., Healthcare versus Finance) 118
6.2.4 Human Cognitive Limitations in Understanding Models 118
6.2.5 Risk of Oversimplification 119
6.3 Methods of Explainability in GenAI 119
6.3.1 Feature Visualization Techniques 119
6.3.2 Model-Agnostic Techniques 120
6.3.3 Intrinsic Explainability Approaches 122
6.3.4 Counterfactual Explanations 122
6.3.5 Example-Based Explanations (Prototypes and Criticisms) 123
6.3.6 Natural Language Explanations and Narrative-Based XAI 123
6.4 Research Aspects in Explainability for GenAI 124
Exercises 125
References 126

Chapter 7 Bridging Fairness and Explainability in AI 128

7.1 The Interplay Between Bias and Opacity 128
7.2 Explainability as a Tool for Fairness Audits 130
7.2.1 Feature Importance Analysis 131
7.2.2 Counterfactual Explanations 131
7.2.3 Algorithmic Debiasing 132
7.2.4 Regulatory Implications 132
7.2.5 Human-in-the-Loop Systems for Accountability 133
7.3 Socio-Technical Perspectives on Fair and Explainable AI 134

7.4	Technical Methods Bridging Fairness and Explainability		135
	7.4.1	Adversarial Debiasing	135
	7.4.2	Re-Weighting and Re-Sampling	136
	7.4.3	Post-Hoc Explanation Tools for Bias Diagnosis	136
	7.4.4	Explainable Fairness-Aware Training Pipelines	136
7.5	Real-World Case Studies		137
	7.5.1	Case Study 1: Hiring Algorithms	137
	7.5.2	Case Study 2: Loan Approval Systems	138
	7.5.3	Case Study 3: Healthcare Diagnostics	139
7.6	Evaluation Metrics and Benchmarking of Fair-Explainable AI Systems		141
Exercises			141
References			142

Chapter 8 Challenges and Limitations 144

8.1	Trade-Offs Between Fairness, Explainability, and Accuracy		144
	8.1.1	The Accuracy	144
	8.1.2	Explainability versus Accuracy Model	145
	8.1.3	Fairness versus Explainability	145
	8.1.4	Multi-Objective Optimization Approaches	146
8.2	Computational Barriers		146
	8.2.1	Resource-Intensive Training	146
	8.2.2	Scalability Challenges	146
	8.2.3	Data Limitations and Quality Constraints	147
	8.2.4	Environmental Sustainability Concerns	147
	8.2.5	Accessibility and Global Divide	147
	8.2.6	Potential Mitigation Strategies	148
8.3	Legal Barriers		148
	8.3.1	Regulatory Fragmentation and Inconsistency	148
	8.3.2	Liability and Accountability	149
	8.3.3	Privacy and Data Protection	149
	8.3.4	Intellectual Property Rights	149
	8.3.5	Algorithmic Bias and Anti-Discrimination Laws	149
	8.3.6	Lack of Explainability in Legal Contexts	150
	8.3.7	Toward AI-Specific Legal Frameworks	150
8.4	Social Barriers		150
	8.4.1	Public Trust and Acceptance	150
	8.4.2	Workforce Displacement and Economic Inequality	151
	8.4.3	Digital Divide and Inequitable Access	151
	8.4.4	Ethical Concerns and Social Acceptance	151
	8.4.5	Cultural and Regional Differences	152
	8.4.6	Media Influence and Public Discourse	152
	8.4.7	Strategies for Overcoming Social Barriers	152

8.5 Standardization Gaps in Evaluation 152
8.5.1 Fragmented Benchmarks and Metrics 153
8.5.2 Lack of Fairness Standardization 153
8.5.3 Explainability and Interpretability Challenges 153
8.5.4 Reproducibility Crisis in AI Research 154
8.5.5 Absence of Cross-Domain Evaluation Frameworks 154
8.5.6 Emerging Standardization Efforts 155
8.5.7 Toward Unified Evaluation Standards 155
8.6 Conclusion and Future Outlook 155
References 156

Chapter 9 Future Directions and Responsible AI 161

9.1 Research Opportunities: Hybrid Models and Symbolic Reasoning 161
9.1.1 The Foundations of Hybrid AI 161
9.1.2 Notable Examples and Industrial Adoption 162
9.1.3 The Role of Symbolic Reasoning in Modern AI 162
9.1.4 Research Outlook 162
9.2 Scalable Solutions for Fairness-Aware Systems 164
9.2.1 Techniques for Debiasing at Scale 164
9.2.2 Key Challenges in Ensuring Fairness in GenAI 164
9.2.3 Tools and Approaches for Scalable Fairness Monitoring 165
9.3 Multidisciplinary Frameworks for Ethical AI 165
9.3.1 Case Example: The Montreal Declaration for Responsible AI 166
9.4 Governance, Policy, and User-Centric Design 168
9.4.1 Global and Regional Policy Frameworks 168
9.4.2 National AI Strategies 168
9.4.3 User-Centric Design Principles 169
9.4.4 Technical Tools for Ethical Implementation 169
9.5 Future Outlook and Open Challenges 169
9.6 Conclusion 171
Exercises 171
References 172

Index 175

Preface

Generative artificial intelligence (GenAI) represents a significant advancement in the evolution of artificial intelligence (AI), enabling machines to generate text, images, audio, video, and code with increasing accuracy and contextual relevance. These capabilities are now widely embedded in educational tools, healthcare systems, the creative industries, and software development environments. As GenAI becomes integral to academic study and professional practice, there is a growing need for structured learning resources that address both its technical foundations and ethical implications.

This textbook is designed to provide a comprehensive and systematic introduction to GenAI, with particular emphasis on fairness, transparency, explainability, and human-centered design principles. While many existing texts focus either on mathematical modeling or on high-level ethical discussions, this book integrates both perspectives to support responsible understanding and application of AI. It recognizes that modern AI practitioners must be equipped not only with algorithmic knowledge but also with the ability to evaluate social impact, bias, accountability, and regulatory considerations.

The book begins by establishing foundational concepts, including the evolution of AI and learning paradigms that support generative modeling. Core architectures, such as Generative Adversarial Networks (GANs), Variational Autoencoders (VAEs), diffusion models, and transformer-based systems, are introduced in a clear and structured manner. Emphasis is placed on conceptual understanding, supported by diagrams, examples, and real-world applications, making the material accessible to students with diverse technical backgrounds and experience levels.

The subsequent chapters focus on the ethical landscape of GenAI. Key topics include the sources of bias in data and algorithms, methods for measuring and mitigating unfairness, and the importance of explainability in complex generative systems. The book examines widely used interpretability techniques and discusses how transparency supports trust, accountability, and compliance with emerging AI governance frameworks. The relationship between fairness and explainability is explored in depth, highlighting their complementary roles in the development of responsible AI.

This textbook is intended for upper-level undergraduate and graduate courses in computer science, AI, data science, and related fields. It is also suitable for educators designing AI ethics curricula and professionals seeking a principled understanding of GenAI systems. Each chapter is structured to support the learning objectives with clear explanations, illustrative examples, and discussion-oriented content.

By combining technical rigor with ethical awareness, this book aims to prepare readers to design, evaluate, and deploy GenAI systems responsibly. This encourages critical thinking about the societal role of AI and promotes the development of technologies that align with fairness, transparency, and enduring human values.

About the Authors

Kirti Seth is a researcher and academician in the field of computer science and engineering with over 20 years of experience in teaching and research. She is currently working as Professor in the School of Computer and Information Engineering at INHA University in Tashkent. She earned her PhD in computer science and engineering in 2016 from the Department of Computer Science and Engineering, Dr. A. P. J. Abdul Kalam Technical University (AKTU), Lucknow, India, with her doctoral research focused on component-based systems. She holds an MTech in computer science (2009) from Banasthali Vidyapeeth, Rajasthan, India, and also possesses an MSc degree in computer science.

Dr. Seth has published more than 50 research papers in reputed international journals and leading publishers and has authored five academic books. Her research work reflects a strong integration of theoretical foundations and applied problem-solving in modern computing systems. She is associated as a subject matter expert (SME) with TCS-iON (unit of Tata Consultancy Services, India).

She has actively organized and participated in seminars, conferences, workshops, expert lectures, and technical events and has conducted training programs for students and faculty members, including her involvement in Google's TechMaker Event (2018). She has delivered keynote addresses at several international conferences. In addition to formal academia, Dr. Seth actively engages in knowledge dissemination through her own YouTube channel, where she teaches computer security and related computer science topics, extending learning beyond the classroom.

Her research interests include component-based systems, service-oriented architecture, bio-inspired optimization techniques, neural networks, artificial intelligence, and computer security. She was awarded the "Young Scientist in Software Engineering" Award in 2017 in recognition of her scholarly contributions.

Nidhi Gupta is Associate Professor in the Department of Computer Science and Engineering, Sharda University, Greater Noida, India. She has over 20 years of academic experience, with a strong emphasis on research and scholarly publishing. Her research interests include artificial intelligence and machine learning, computer vision, and information retrieval. She has authored several research articles in reputed international journals and conferences and has contributed to edited books in computer science.

Ashish Seth is a teacher, a consultant, a researcher, and an author. He is Professor at the School of Computer and Information Engineering, Inha University in Tashkent, Uzbekistan. He is a senior member of Institute of Electrical and Electronics Engineer (IEEE), Association of Computing Machinery (ACM), and Computer Society of India (CSI). He also served as ACM-distinguished speaker. He has a PhD (computer science) in the area of information systems from Punjabi University, Patiala, Punjab, India, and holds an MPhil (CS), MTech (IT), and MCA degrees. He has published more than 50 research papers in indexed journals. He has authored five books and several book chapters. He also edited two books and served as an editor of an indexed journal. He has been into research and academics for more than 22 years. He has worked at various universities in India and abroad, holding different positions and responsibilities. He has organized and participated very actively in various conferences, workshops, and seminars.

He is associated as a subject matter expert (SME) with TCS-iON (a unit of Tata Consultancy Services, India), as an examiner with IBIS (International Baccalaureate Information System), and as a mentor of Change for Atal Innovation Mission (an initiative of NITI Aayog, Government of India). He has been a consultant to some projects with Aziksa, Erasmus, and UNDP Uzbekistan.

His work bridges traditional academic instruction with modern digital resources, fostering expertise across various technological domains. His research interests include service-oriented architecture, cloud computing, AI, and blockchain technologies. He finds interest in reading and writing articles on emerging technologies.

Abbreviations

ANN	Artificial Neural Network
CNN	Convolutional Neural Network
DAE	Denoising Autoencoder
DL	Deep Learning
FFN	Feed-Forward Neural Network
GAN	Generative Adversarial Network
GenAI	Generative Artificial Intelligence
GPU	Graphics Processing Unit
GRU	Gated Recurrent Unit
LLM	Large Language Models
LSTM	Long Short-Term Memory
ML	Machine Learning
MSE	Mean Squared Error
NLP	Natural Language Processing
PCA	Principal Component Analysis
ReLU	Rectified Linear Unit
RNN	Recurrent Neural Network
SVM	Support vector machines
VAE	Variational Autoencoders
XAI	Explainable AI

1 Foundations of Generative AI

1.1 AI AND ITS EVOLUTION

The artificial intelligence (AI) systems aim to exhibit human intelligence that is autonomous and adapts to changing environments. The history of Generative AI (GenAI) is not very long. However, machine learning (ML) and deep learning (DL)-based systems are the foundation stone of GenAI. Over the years, the technology has advanced and matured into systems that can generate new text, audio, images, and video. This generation of new, meaningful content is termed "Generative AI."

The advancements in this field started from the 1950s when Arthur Samuel developed an algorithm for playing chess. During this period, the AI systems were rule-based, with symbolic AI and expert systems. MYCIN for medical diagnosis and DENDRAL in the chemistry domain were some of the well-known expert systems developed at that time; however, they lack scalability, have ambiguity, and poor capacity to handle uncertainties (Hu et al., 2009). In 1950, the Turing test was proposed by Alan Turing, which laid the foundation of machine learning. At the end of the 20th century, it was realized that AI is not only about rule-based processing but also about learning from examples and adapting in new environments. This view gave rise to the concept of "machine learning."

A significant transformation in AI systems took place in the 21st century. The explosion of digital information took place through the advancements in the internet, sensors, and smartphones. Digital advancement, including images, text, audio, and videos, has created a big data environment that provides large datasets for training AI systems. It enables automatic feature extraction and the use of sophisticated DL-based models. The revolution in specialized processors Graphics Processing Units (GPUs) enabled parallelized computation in order to overcome the need for high computation and processing power (Dally et al., 2021). The evolution of AI system to GenAI is shown in Figure 1.1.

A notable enhancement in this research area was made with the invention of complex DL-based algorithms such as Convolutional Neural Networks (CNNs) and Recurrent Neural Networks (RNNs) for solving computer vision applications and natural language processing (Saon & Picheny, 2017).

The revolution came with the advent of GenAI. In contrast to the algorithmic development that was based on classification or prediction, the GenAI systems aim to produce new data in the form of images, text, audio, and video. Table 1.1 summarizes the evolution of AI from rule-based systems to GenAI.

A noteworthy progression is seen in GenAI applications. Therefore, the concepts of ensuring ethics, trust, and responsibility remain equally important. The next section discusses the fairness and explainability associated with GenAI.

DOI: 10.1201/9781003713951-1

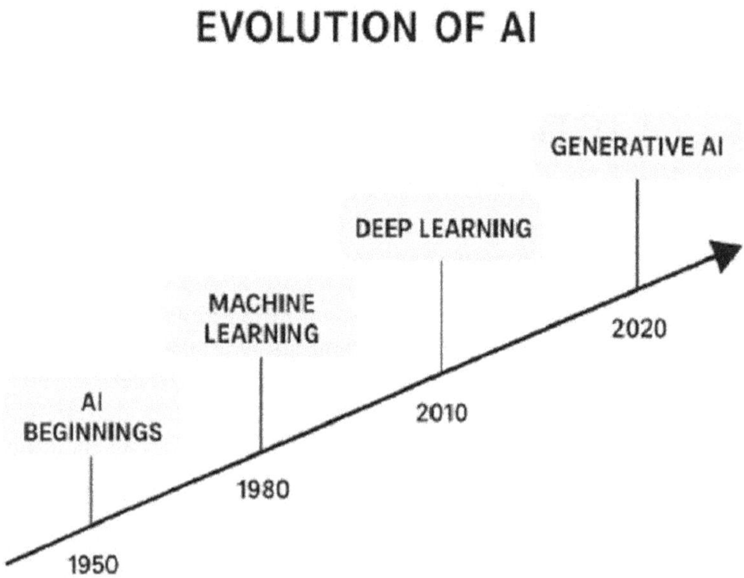

FIGURE 1.1 AI Evolution to Generative Systems

TABLE 1.1
Evolution of AI

Era/Paradigm	Approx. Period	Key Features	Examples	Limitations
Rule-Based Systems (Symbolic AI)	1950s–1980s	Logic-driven, knowledge based, decision trees	MYCIN (medical diagnosis), DENDRAL (chemistry)	Inflexible, poor handling of uncertainty, limited scalability
Traditional Machine Learning	1980s–2000s	Data-driven, statistical models, feature engineering	Decision trees, SVMs, Bayesian networks	Dependent on manual features, moderate scalability
Deep Learning	2010s onwards	Multilayer neural networks, representation learning, GPU/TPU acceleration	AlexNet (2012), AlphaGo (2016), ResNet	Requires large datasets, high computational cost, black-box nature
Generative AI	2014 onwards	Content creation, self-attention mechanisms, multimodal learning	GANs (2014), GPT models, Stable Diffusion, DALL·E Series	Ethical risks, data biases, computationally intensive

1.2 UNDERSTANDING GENERATIVE AI, FAIRNESS, AND EXPLAINABILITY

One of the most significant developments in computer science in recent years has been GenAI. Generative models are focused on producing new data that closely resembles human-generated content, in contrast to conventional AI systems, which typically seek to categorize, predict, or recommend based on historical data. The distinction between creativity and automation is blurred by these models' ability to produce realistic images, cogent text, synthetic audio, and even video sequences. Generative models such as Generative Adversarial Networks (GANs), Variational Autoencoders (VAEs), and transformer-based models like GPT and DALL·E have transformed a variety of industries, including healthcare, design, and education (Bengesi et al., 2024).

At its core, GenAI discovers the more complex patterns and structures in large datasets and uses them to produce outputs that appear novel but authentic. For example, a language-modeling system trained on varied corpora of language can generate human-like essays, while a picture-producing model can generate completely novel faces or pieces of art. This capability has unleashed vast possibilities in content generation, drug discovery, personalized learning, and scientific simulations, but it has also posed challenges for misuse, including deepfakes, disinformation, and copyright issues (Sharma & Sharma, 2024). Consequently, the GenAI conversation cannot be decoupled from concerns about fairness, transparency, and accountability.

1.2.1 Fairness in GenAI

Fairness in AI is the concept that algorithmic systems should function without bias or discriminatory injustice (Xiang, 2024). Since GenAI models are trained on large datasets drawn from the internet or other stores, they are prone to inheriting biases contained in the datasets (Choudhry et al., 2024). For instance, a model for generating text might replicate stereotypes or an image generator might underrepresent particular demographic groups. It becomes important to address as it may cause social harm or inequalities.

Achieving fairness entails both technical and moral considerations. Technically, tools like balanced dataset creation, bias identification, and debiasing algorithms are under development. Ethically, it calls for consideration of whose voices and opinions are included in training data and whose are excluded. GenAI fairness is not an issue of accuracy—it is an issue of trust, equity, and inclusion in the form that AI technologies take society.

1.2.2 Explainability in GenAI

Fairness is closely connected to the idea of explainability. Explainability means making AI systems intelligible to humans, not only how they come to decisions but also why they generate certain outputs. Rule-based systems were by definition interpretable because their decisions were based on explicit rules. However, recent generative models, particularly deep learning systems with billions of parameters, tend to be "black boxes." Their decision-making mechanisms are intricate, nonlinear, and hard to follow (Schneider, 2024).

Not able to explain the decision is risky in high-consequence domains like medicine, finance, and the law, where the answer of "why" is much important instead of result. In GenAI, explainability can assist in establishing whether a model is generating biased content, if it can be relied upon to support professional decision-making, and how to make it more reliable. Methods like attention visualization, model probing, and post-hoc explanation frameworks are being proposed to increase transparency, though this continues to be an emerging area of study.

1.2.3 The Interconnection of GenAI, Fairness, and Explainability

GenAI's impact on society is determined not only by its technical capabilities but also by how fairly and transparently it operates. Fairness ensures that outputs do not reinforce harmful stereotypes or discrimination, while explainability ensures that stakeholders understand and trust the technology. Together, they form the ethical backbone of responsible AI development.

Students, researchers, policymakers, and industry practitioners must thus have a deep understanding of generative AI, fairness, and explainability. Due to its widespread applications, GenAI has become increasingly integral to everyday life. Ensuring a creative and accountable and fair system will be one of the hallmark challenges of the AI age.

1.3 HISTORICAL DEVELOPMENT OF GENAI

The journey of GenAI is not very long but the roots of GenAI was established in the 1950s when Alan Turing introduced Turing Test for exhibiting the machine intelligence. The evolution of GenAI is closely originated from the history of AI. The Turing Test embarks further investigation in the domain of cognitive science, philosophy, linguistic, and scientific studies. This give rise to the concept of ML that is a very significant basis for the development of generative systems (He et al., 2025).

A basic neural model, Perceptron, was first introduced by Frank Rosenblatt in 1957 that was one of the first of the neural network models (Estebon, 1997). It acts as the beginning of exhibiting ML's capability to simulate human brain functions. The concept of backpropagation was introduced further in the 1980s that witnessed a giant leap in training deep neural networks. The concept of neural networks and DL become the foundations of generative models later on. One of the significant achievements in AI was seen in 1997 when IBM's Deep Blue's win over world chess champion Garry Kasparov (Bory, 2019). It also highlights the limitation of rule-based systems as well as the necessity for data-driven, adaptive intelligence. Table 1.2 summarizes some of the major advancements in the field of AI.

The year 2006 showcases another big achievement when Geoffrey Hinton and others developed a deep belief network, enabling the training of complex architectures (Hinton et al., 2006). It revived DL through the development of deep belief networks and helped in the evolution of generative systems. But it was in 2014

TABLE 1.2
Advancements in AI

Year	Advancement
1950	Alan Turing proposes the Turing Test, thereby laying the conceptual foundation for machine intelligence.
1957	Perceptron—an early neural network model—introduced by Frank Rosenblatt.
1980s	Rise of neural networks and backpropagation for learning complex patterns.
1997	IBM's Deep Blue defeats world chess champion, showcasing symbolic AI strength.
2006	Geoffrey Hinton popularizes deep learning through deep belief networks.
2014	GANs—a major breakthrough in generative modeling.
2017	Google introduces Transformer architecture (Attention is All You Need), foundational to modern GenAI.
2018	GPT-1 (OpenAI) launches the first major transformer-based language model for text generation.
2019	GPT-2 released with 1.5 billion parameters, capable of realistic text generation.
2020	DALL·E and CLIP (OpenAI) announced, bringing GenAI to image generation and multimodal understanding.
2020	GPT-3 released with 175 billion parameters, enabling zero-shot and few-shot learning capabilities.
2021	Codex powers GitHub Copilot, bringing GenAI into code generation.
2022	Stable Diffusion, Midjourney, and DALL·E 2 popularize GenAI art creation.
2022	ChatGPT released (based on GPT-3.5) the first viral conversational GenAI application.
2023	GPT-4 and Claude launched, demonstrating improved reasoning and multimodal capabilities.
2024	Widespread adoption of custom GenAI, fine-tuned models, and AI copilots across industries.
2025	Ongoing progress in multimodal AI, video generation, agent-based GenAI, and real-time deployment.

that GenAI began to take form with the development of GANs by Ian Goodfellow (Goodfellow et al., 2014), which enabled machines to create realistic images and media using adversarial learning. The transformation in text generation was achieved in 2017, when Google researchers developed the Transformer architecture—a model entirely dependent on self-attention mechanisms, which was extremely successful for sequence understanding and generation. (Vaswani et al., 2017).

After the introduction to Transformer, OpenAI released its first major release in 2018 with GPT-1, beginning the age of Generative Pre-trained Transformers. A year later, GPT-2 demonstrated the capabilities of Large Language Models (1.5 billion parameters) to generate text that is comparable to human content, attracting attention from the public to AI-generated content. In 2020, OpenAI extended even further with GPT-3, which possessed 175 billion parameters and showed the ability to accomplish tasks based on few or no training samples (Zong & Krishnamachari, 2022). OpenAI introduced DALL·E and CLIP, extending GenAI into the multimodal realm, enabling image generation and text-image understanding. In 2021, Codex, a derivative of GPT-3, came out, which powered GitHub Copilot, a program that could write and assist with coding code in response to natural language requests. And then 2022 saw a major change in GenAI, with Stable Diffusion, Midjourney, and DALL·E 2 delivering high-quality, easy-to-use AI-created art (Blattmann et al., 2023). The same year, ChatGPT was released, providing conversational access to GPT-3.5, which went on to become an instant world phenomenon owing to its natural, intelligent answers, and universal applicability. The timeline of AI advancement in different time periods is shown in Figure 1.2.

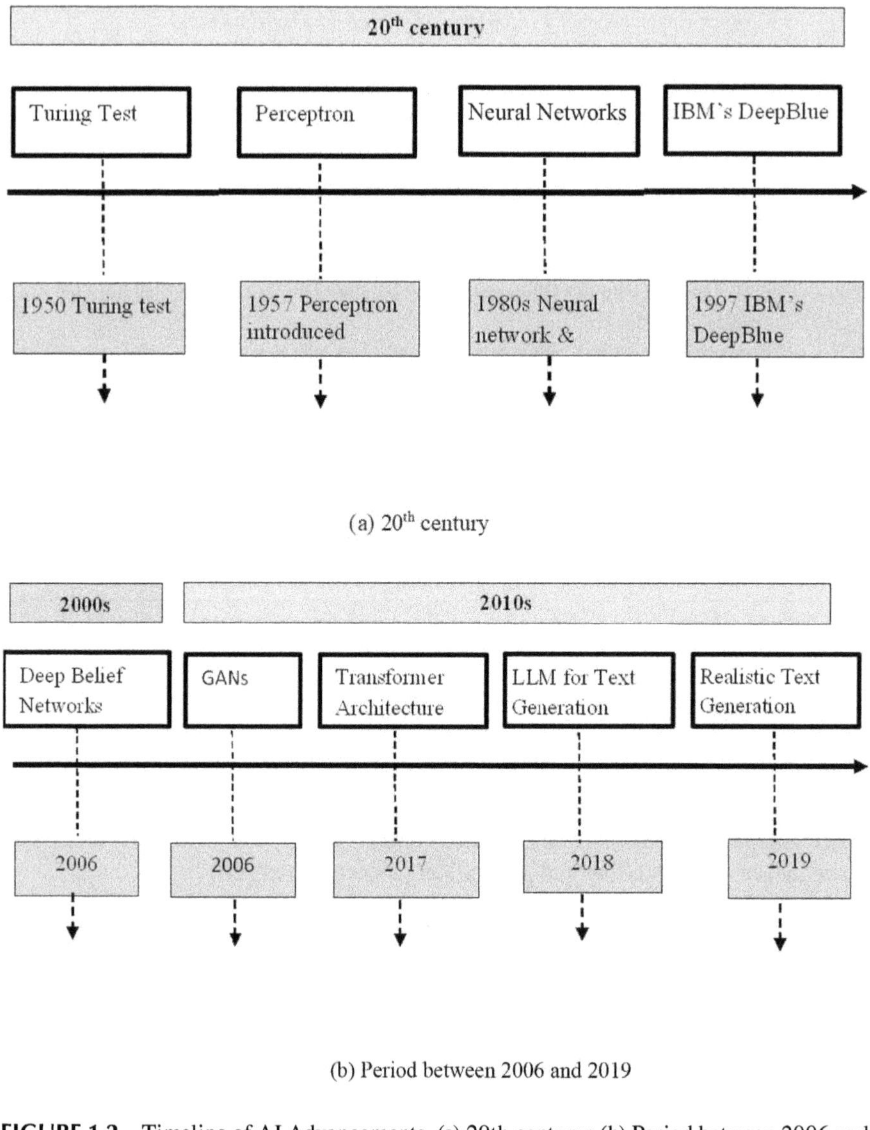

FIGURE 1.2 Timeline of AI Advancements. (a) 20th century; (b) Period between 2006 and 2019; (c) Period between 2020 and 2025

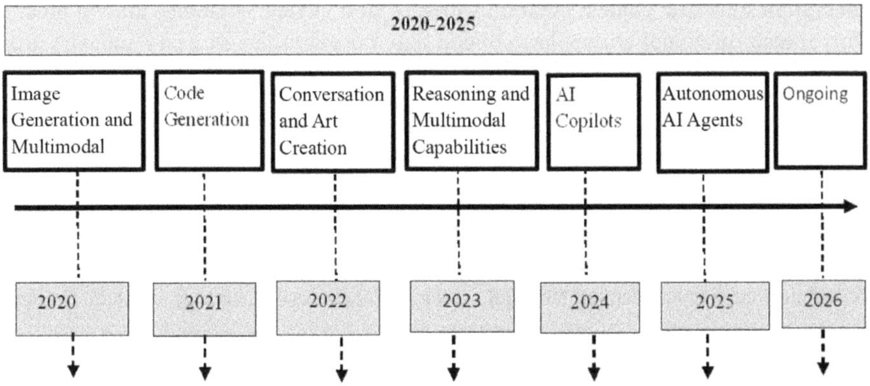

(c) Period between 2020 and 2025

FIGURE 1.2 (Continued)

By 2023, GPT-4 came with better reasoning, multimodal input (text and image), and safer alignment, and rivals such as Claude from Anthropic brought other GenAI platforms. By 2024, companies started adopting custom fine-tuned GenAI models for domain-specific use, and AI copilots were becoming part of productivity software, health systems, and business applications. The rise of agentic systems was introduced in 2025. It made AI agents to behave intelligently, and independently. As we approached 2026, GenAI is progressing with innovation in video creation, real-time agents, and multimodal assistants and changing sectors and how humans engage with smart systems.

1.4 AI AND GENAI

AI is a branch of computer science that aims to develop systems capable of accomplishing tasks that were formerly deemed to require human intelligence. These include sectors like machine learning (learning from experience), natural language processing (analyzing and processing human language), and computer vision (perceiving the environment). Fundamentally, AI aims to mimic and enhance human intellectual capabilities—for example, reasoning, problem-solving, decision-making—so machines can act independently or aid humans in intricate tasks.

GenAI is a subset under AI that deals with models that can produce new, unique data based on its learning. These models do not merely analyze or classify stuff, rather they create it. Once given a prompt, GenAI models can generate coherent text, photorealistic images, human-like audio, code for computers, or even whole videos. Real-world uses of GenAI are growing fast and include software such as conversational bots (chatbots), auto-translators of language, AI-powered art and music generation software, content creation software, and virtual assistants, among others. This transformational ability has created new opportunities in every industry that may be education and medicine, or entertainment and design.

1.5 AI PILLARS: ML, DL, GENAI, AND LLM

AI refers to making machines intelligent. It involves various approaches that learn machines to behave autonomously, making decisions or adapt to the changing environment. "Artificial intelligence" is a broad term. It encompasses many other AI-related concepts such as ML, DL, GenAI, and large language models (LLM). Figure 1.3 shows the relationship between all these concepts. A brief introduction of other key terms is provided in the following sections.

1.5.1 Machine Learning

ML is all about allowing systems to learn patterns within data and predict or decide things based on that, all without being programmed to do so. Tasks under it are classification, regression, and clustering.

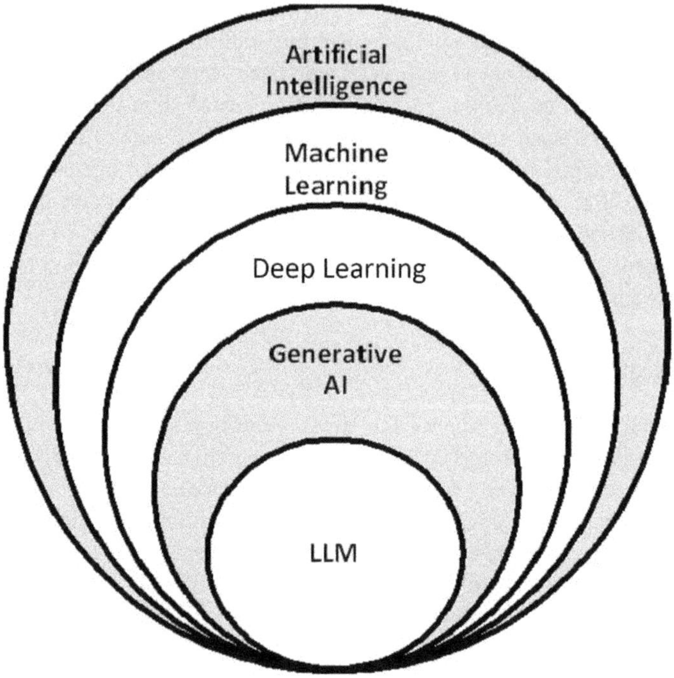

FIGURE 1.3 Diagrammatic Representation of Relationship Between AI, ML, DL, GenAI, and LLM

ML and GenAI go hand in hand but both serve different purposes while using in context of AI. To compare, GenAI is a category of ML that has been specifically developed to generate new content such as text, images, music, or code, after deep analysis of underlying pattern of given data. While all GenAI models are based on ML, all ML models are not generative. That is, ML is all about learning and predicting, but GenAI is all about producing.

1.5.2 Deep Learning

Both DL and GenAI are related notions in AI, but they vary in emphasis and use. DL originates from ML that employs neural networks with more than one layer to learn patterns automatically from large datasets. (Wang et al., 2016). DL drives most AI systems, such as those for image recognition, voice processing, and self-driving cars. GenAI, in contrast, is an AI model that is frequently developed with DL, specifically created to produce new content, like text, images, audio, or code. DL essentially gives the underlying architecture (such as transformers or GANs) that makes GenAI possible. Thus, DL is the *technology*, but GenAI is an application or use case of that technology, with the aim of *creation*.

1.5.3 Large Language Models

LLMs are a class of advanced AI models trained on massive amounts of textual data to understand, interpret, and generate human language. LLMs are built using DL architectures like Transformers, and they can perform a wide range of language-related tasks such as answering questions, summarizing text, translating languages, generating code, and more. Examples include OpenAI's GPT series, Google's Gemini, and Meta's Llama models (Chang et al., 2024).

LLMs are a subset of GenAI that specializes in the processing and generation of human language. GenAI is the larger umbrella under which models are included to generate all kinds of content such as images, audio, video, and text. Although all LLMs qualify as GenAI, not every model of GenAI is an LLM. For example, generative image models such as DALL·E or generative music composition models are instances of GenAI but not LLMs. Essentially, LLMs are language-oriented specialists, while GenAI is a cross-modal content approach.

1.6 GENAI BLUEPRINT

The methodology adopted by GenAI involves a systematic pipeline that transforms raw input data into newly generated content through a series of well-defined stages. This process combines foundational techniques from ML, particularly DL, and leverages powerful generative models to create novel outputs that resemble the training data. The typical stages of this methodology are shown in Figure 1.4, and each stage is described in the following sections.

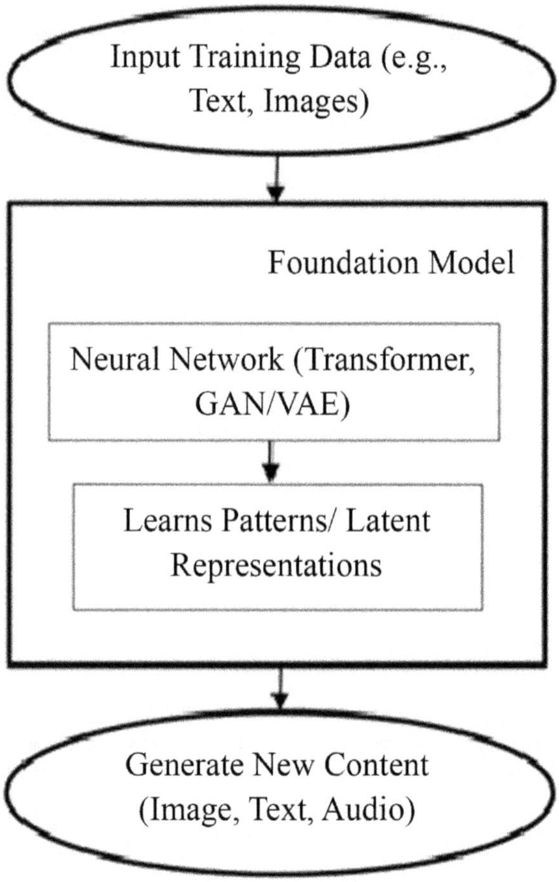

FIGURE 1.4 Block Diagram of Generative AI Basic Procedure

1.6.1 Input Training Data

The GenAI process gathers and processes large volumes of data that serve as the foundational knowledge base for the generative model. This data can be textual, visual, auditory, or multimodal, depending on the application domain. High-quality, diverse datasets are crucial for enabling the model to learn rich and meaningful representations.

1.6.2 Foundation Model Training

At the core of GenAI lies the foundation model. This model is a large, pre-trained neural network capable of capturing complex patterns in the data. This model is trained using advanced architectures such as Transformers, GANs, or VAEs.

1.6.3 Generation Phase

Once the model is trained, it can be used to generate new content by sampling from the learned latent space. The latent variables serve as the internal representation from which the model constructs novel outputs. Depending on the model type and training objective, the generated content can range from realistic images and coherent text to music, code, and more.

The GenAI pipeline involves training on large datasets using deep neural network models, learning latent representations, and using those representations to synthesize new, high-quality content. This methodology underpins the remarkable capabilities of modern generative systems, such as ChatGPT, DALL·E, and other foundation models.

Understanding the fundamental ideas behind the development of GenAI is crucial to realizing this vision. These fundamental components of GenAI are depicted in Figure 1.5. The key foundation areas of understanding Generative System are related to the following:

- *Mathematics*: Mathematics covers the core mathematical foundations required for understanding of GenAI algorithms.
- *Models*: The models provide the details of various algorithms utilized for generating new contents.
- *Evaluation Metrics*: The metrics defines various evaluation parameters used for evaluating the performance of different algorithms.
- *Training*: It defines different optimizations applied for training the GenAI model.
- *Ethics*: This explains the various ethical requirements for ensuring trust, fairness, and explainability of algorithmic results.

The forthcoming sections provide detailed explanation of each key foundation.

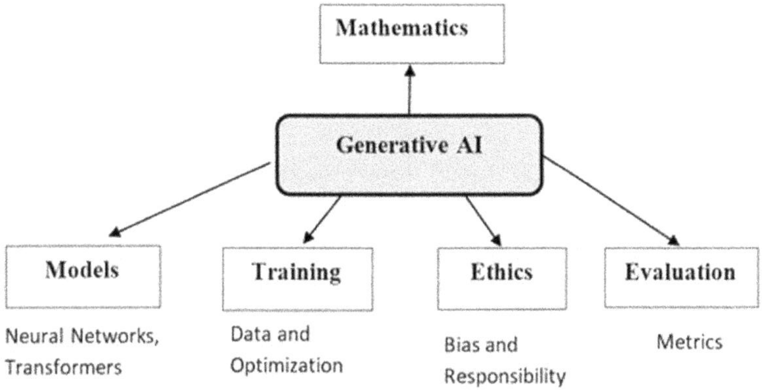

FIGURE 1.5 Foundation Areas of GenAI

1.7 CORE FOUNDATIONS OF GENAI: MATHEMATICS, MODELS, AND METRICS

1. Mathematical foundations
2. Foundation models of ML and DL relevant to GenAI
3. GenAI engine: Transformers
4. Model loss function
5. Evaluation metrics

1.7.1 Mathematical Foundations

1.7.1.1 Basics of Probability Theory

Probability theory is the mathematical foundation of most models used in GenAI. Probability theory is essential in predicting, generating the content using GenAI. The model produces the outcome after learning the probability distributions. A probability distribution indicates the likelihood of each possible outcome in the given scenario.

The key terms used in probability are as under:

- *Random Variables*: A random variable is a value that is determined by random chance. In generative modeling to generate images, random variables represent pixel values or other image features; to generate text, random variables represent the probability of each word coming next in a sentence; and to generate music, random variables could be used to represent notes, duration, and volume.
- *Joint Probability*: Joint probability describes the likelihood of two or more events that occur simultaneously. Suppose XX and YY are the two discrete events, the joint probability is defined as shown below:

$$P(XX,YY) = P(XX = x, YY = y)$$

It represents how likely that event *XX* takes the value x and event *YY* takes the value as y.

If *XX* and *YY* are continuous variables, then joint probability density function is defined as function $f(x, y)$ such that:

$$f(x,y) \geq 0, \text{for all}(x,y) \in \mathbb{R}$$

$$\int\int_{-\infty}^{+\infty} f(x,y)\,dx\,dy = 1$$

For any set $A \in \mathbb{R}^2$:

$$(P(XX,YY) \in A) = \int\int_{-\infty}^{+\infty} f(x,y)\,dx\,dy$$

For generative models, joint probability is important as data elements tend to be interrelated. For example, in case of language generation, the next word in a sentence depends on its predecessors. The joint probability encodes these dependencies so that models can produce text that makes sense instead of random words. VAEs model the joint distribution *P(x,y)* where *XX* is the observed data (e.g., image), and *YY* is a latent variable. The GANs implicitly learn the joint distribution between the data and the generator output.

- *Conditional Probability*: The conditional probability describes the likelihood of occurrence of one event when another event has already happened. It is denoted by *P(XX / YY)*. It indicates the probability of occurrence of event *XX* when *YY* has already occurred.

  In the context of image generation, the conditional probability is used to complete an incomplete image, Similarly, the word in the sentence can be determined based on the probability of next word.

  Bayes' Theorem: The Bayes' theorem defines the conditional probability as follows:

$$P(XX / YY) = \frac{P(YY / XX)P(XX)}{P(YY)}$$

where:

P(XX/YY) is the probability of occurrence of event *XX* given *YY*.
P(YY/XX) is the probability of occurrence of event *YY* given *XX*.
P(XX) is the probability of event *XX*.
P(YY) is the probability of event *YY*.

In GenAI, many models such as VAE rely on learning *P(X/Z)* where the probability of generating data *X* given latent variable *Z*.

- *Marginal Probability Distribution*: Marginal probability deals with the probability of occurrence of single event, while ignoring the other events. If the event *XX* and *YY* are discrete, then marginal probability is computed using the equation as follows:

$$P(XX) = \sum P(XX, YY)$$

If the event *XX* and *YY* are continuous, then marginal probability is defined by integrating the probabilities as shown:

$$P(XX) = \int P(XX, YY)\,dy$$

1.7.1.2 Relevance of Probability Theory in GenAI

There are several ways probability theory can be used in GenAI. Table 1.3 lists some of the approaches that are based on using probability theory.

TABLE 1.3
Probability Theory in GenAI

Model Type	Uses of Probability Theory
VAEs	Approximate joint, conditional, and marginal distributions
GANs	Implicit joint distribution via adversarial training
Diffusion Models	Conditional distributions of noise and data across timesteps
Bayesian Models	Prior, posterior, and likelihood estimation using conditional/marginal probabilities

1.7.2 Optimization

Optimization is a mathematical problem to maximize or minimize any quantity. In ML, optimization is performed to minimize the error in training the model and thereby improving the model accuracy.

The optimization function is a measure of the goodness of fit of the model in numerical terms. The lower the loss function, the better the model fits the training data.

Optimization algorithms (e.g., Gradient Descent, Adam, RMSProp) apply the gradient of the loss function with respect to model parameters to decide the direction and magnitude of the updates for parameters during the training procedure. The model 'learns' by iteratively adjusting its parameters in the direction that minimizes the output of the optimization function, until the function's value is minimized below a certain threshold. This iterative tuning is to bring the model to a position of minimum loss.

For example, the optimization functions applied in regression problems are mean squared error (MSE). It estimates the average of the squared error of predicted and actual values. Cross-Entropy Loss is optimization function applied in classification problems. It is used to estimate the difference between two probability distributions (predicted class probability and true class probability).

While the optimization function is minimized on training data, the real goal is to have good performance on new, unseen data (generalization). The optimization function and algorithm selection are responsible for attaining this.

1.7.3 Foundation Models of ML and DL Relevant to GenAI

1.7.3.1 Neural Networks

1.7.3.1.1 Artificial Neural Network, Activation Functions

Artificial neural network (ANN) is a mathematical model inspired by the human brain. It is composed of several nodes interconnected to form a network. A Multi-Layer Perceptron is the simplest form of a neural network architecture. It consists of three layers: Input layer, Hidden layer, and Output layer. The Input layers accept the input(features) to the network, the Middle layer is called Hidden layers that process the inputs using weighted connections, and Output layer produces the final prediction or classification.

In the ANN when every neuron in a layer is linked to all neurons in the subsequent layer this is a called fully connected layer or fully connected neural network. Activation functions are mathematical formulas that decide the output of a node in a neural network. They add nonlinearity to the model, allowing the network to learn about complicated patterns in the data, required to do tasks such as image recognition and language processing.

1.7.3.1.2 Backpropagation

Backpropagation is the learning algorithm employed to train neural networks. It modifies the network weights to minimize the discrepancy between the predicted and actual output. Backpropagation is the training algorithm for neural networks. It operates in two steps:

- *Forward Pass*: Input is propagated through the network layer by layer. The output produced is compared with actual label and loss function (e.g., MSE, cross-entropy loss) is computed.
- *Backward Pass*: Perform error/loss calculation and weights are updated using gradient descent method iteratively to minimize the loss function.

Table 1.4 shows the basic concepts of neural network with their formula.

1.7.3.1.3 Convolutional Neural Networks (CNNs)

CNNs are a type of deep neural network designed to handle and analyze grid-structured data, such as images. Unlike standard fully connected neural networks, CNNs have a specialized architecture that leverages the spatial nature of input data so that they can extract hierarchical features efficiently and scalable. Figure 1.6 shows the architecture of a CNN.

CNNs were initially created for image classification and object detection, yet their strong capability to extract features has enabled them to be a core building block in GenAI, particularly for applications like image synthesis, image translation, and super-resolution.

TABLE 1.4
Basic Concept of Neural Network

Concept	Purpose	Formula
Activation Function	Introduces nonlinearity	ReLu: $f(x) = \max(0,x)$
Loss Function	Measures prediction error	MSE: $= \frac{1}{n}\sum(y - y')^2$
Gradient	Direction of steepest descent	$\frac{\partial \text{Loss}}{\partial w}$
Weight Update	Adjust weights to reduce error	$w = w - \eta \cdot \text{gradient}$

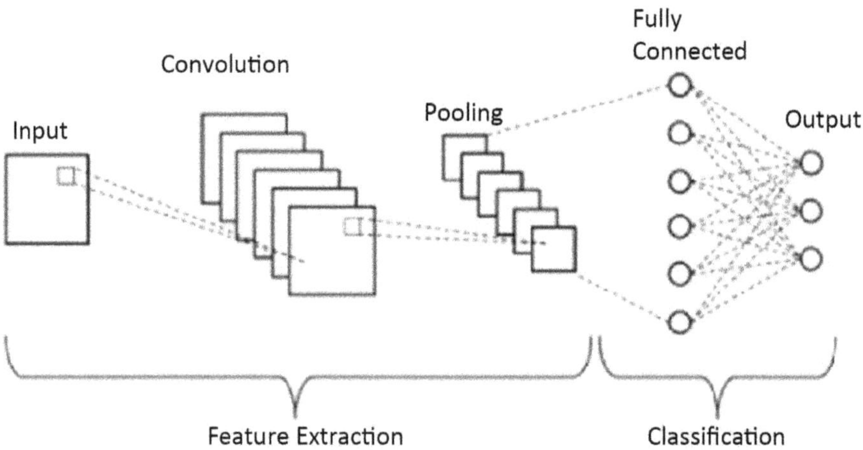

FIGURE 1.6 CNN Architecture

Source: Rhee (2018)

Important elements of CNNs are as follows:

- *Convolutional Layers*: The fundamental building block of a CNN is the convolutional layer, which convolves the input image using a collection of learnable filters (kernels). The filter sweeps across the image in space, generating a feature map that identifies specific patterns like edges, textures, or shapes.
    - *Mathematical operation*: Convolution operation involves moving the filter over the input and calculating dot products.
    - *Receptive field*: Filters capture local features that are subsequently combined to produce global representations.
- *Activation Functions*: Following convolution, an activation function is used to add nonlinearity. Rectified Linear Unit (ReLU) activation function is the most widely used function in CNNs.
- *Pooling Layers*: Pooling layers decrease the spatial size (width and height) of the feature maps, serving to reduce computational requirements and prevent overfitting.
- *Fully Connected Layers*: These are usually placed at the end of the network in the case of normal classification problems. However, in the case of generative models, fully connected layers are replaced with transposed convolutions layers.

1.7.4 Autoencoders

Autoencoders are an ANN architecture that learns compact data representations (encodings) in an unsupervised framework. Their main objective is data

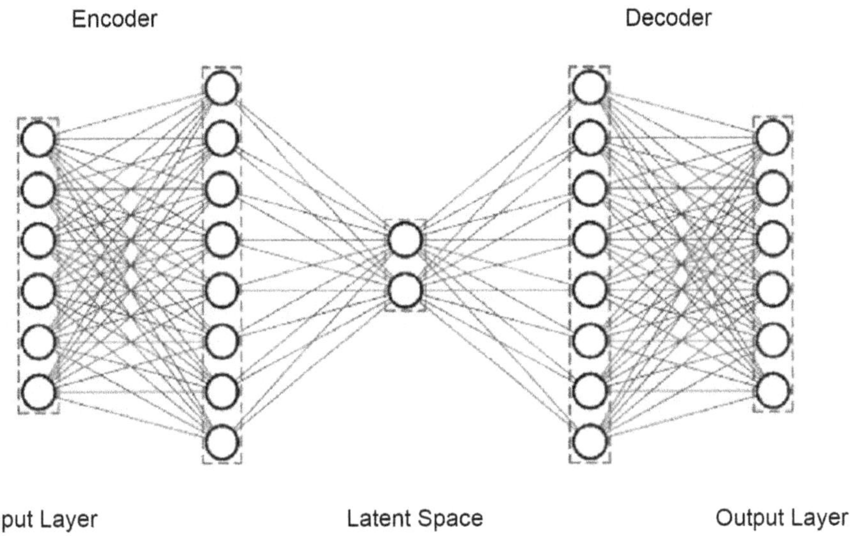

FIGURE 1.7 Representation of Autoencoder Architecture Diagram

Source: Petrov and Wortmann (2021)

reconstruction and compression—they reduce the input data into a compressed low-dimensional form (encoding) and restore it to the original representation (decoding).

An autoencoder is comprised of three important parts. Figure 1.7 provides the architecture diagram of autoencoders indicating their components. The details of its components are as follows:

- *Encoder*—Squeezes the input into a latent-space representation.
- *Bottleneck*—The compact representation (latent vector) with most informative information.
- *Decoder*—Inverts the latent representation to get the input.

Autoencoders are employed extensively in dimensionality reduction, denoising, anomaly detection, image compression, and generative modeling.

Types of Autoencoders

- *Undercomplete Autoencoder*: The latent space has less dimension than the input. This model forces the network to discover the most significant features. It is often used for dimensionality reduction (like Principal Component Analysis (PCA) but nonlinear).
- *Sparse Autoencoder*: It imposes a sparsity constraint on the latent representation. It encourages only a few neurons to be active simultaneously,

picking up more meaningful patterns. This type of model is valuable for performing feature extraction.

- *Denoising Autoencoder (DAE)*: DAE is trained to reconstruct the original input from a noisy version of it. It increases robustness of learned features. It is widely applied in noise removal, image restoration, and pretraining.
- *Contractive Autoencoder*: It includes a penalty in the loss function to make the encoder immune to small changes in the input. It aids in learning invariant and stable features.
- *Variational Autoencoder (VAE)*: It is a probabilistic autoencoder. It learns the distribution over the latent space, instead of learning a constant latent vector. It is widely utilized in generative modeling to generate new samples of data.
- *Convolutional Autoencoder*: It utilizes convolutional layers instead of fully connected layers for model training and development. It is particularly suitable for image data, learning spatial hierarchies.

Autoencoders bridge the gap between feature learning and data compression. There are expert variations in autoencoders that are dedicated to removing noise, generative modeling, and enhanced representation learning.

1.7.5 Recurrent Neural Networks, LSTMs, and GRUs

1.7.5.1 Recurrent Neural Networks

RNNs are a category of neural networks that have been made to handle sequential data by maintaining a hidden state which carries information from past inputs. In contrast to feedforward networks, RNNs contain loops that enable information to flow through from one time step to another. This makes them well suited for applications such as speech recognition, time-series prediction, language modeling, and text generation.

The traditional RNN suffers from the vanishing and exploding gradient problem due to which they are used in limited applications that requires learning long-term dependencies. To overcome this issue, Long Short-Term Memory (LSTM) comes into the picture.

1.7.5.2 Long Short-Term Memory

LSTMs are a newer form of RNNs that is intended to solve the vanishing gradient problem. They provide a memory cell and three gates:

- *Forget Gate*—Determines what information to forget.
- *Input Gate*—Determines what new information to remember.
- *Output Gate*—Determines what information to output.

These gates enable LSTMs to effectively preserve and update long-term dependencies. LSTMs find broad applications in machine translation, text generation, speech synthesis, and sequence prediction.

1.7.5.3 Gated Recurrent Units

Gated Recurrent Units (GRUs) are a reduced form of LSTMs that reduce the problem of vanishing gradients. They merge the forget and input gates into one update gate. In contrast to LSTM, GRU has only two gates to store and manage the flow of information, that is, update and reset gate. It also combines the memory state with the hidden state.

Important components of GRU are:

- *Update Gate*—Regulates how much previous information to retain.
- *Reset Gate*—Determines how much previous information to erase when computing the new state.

GRUs take fewer parameters than LSTMs; hence, they are quicker to train without compromising on dealing with long-term dependencies. They are typically used when there is a need for computational efficiency.

1.8 GENAI ENGINE: TRANSFORMERS

Transformers are a foundational architecture in modern DL, particularly for natural language processing (NLP), machine translation, and generative AI models. They are introduced by Vaswani et al. (2017). In comparison to traditional recurrent and convolutional operations, the Transformer has a special feature called attention mechanism. It emphasizes on the most relevant information in the given input sequence, enabling parallelization and improved long-range dependency modeling. The attention mechanism computes the weights for each input sequence, which determines the most priority and focused text with respect to the given task.

The main components of a Transformer are as follows:

- *Encoder*: The encoder processes the input sequence and converts it into a set of continuous representations. Each encoder consists of multiple identical layers, and each layer comprises of the following:
  - *Multi-Head Self-Attention Mechanism*: It splits the input to multiple heads and captures contextual relationships by performing multiple parallel attention calculations.
  - *Feed-Forward Neural Network (FFN)*: It is a position-wise fully connected network applied after attention.
  - *Add and Norm Layers*: Residual connections and layer normalization are applied to stabilize training.
  - *Positional Encodings*: They are added to the input embeddings to provide sequence order information as the transformer lacks recurrence.
- *Decoder*: The decoder generates the output sequence step by step, using previously generated tokens and encoder outputs. Each decoder layer contains:
  - *Masked Multi-Head Self-Attention*: It prevents positions from attending to future tokens, ensuring autoregressive generation.

- *Encoder–Decoder Attention*: It attends to the encoder's output, enabling the decoder to focus on relevant parts of the input sequence.
- *Feed-Forward Neural Network and Add and Norm Layers*: They are similar to the encoder.

1.8.1 Attention Mechanism

The core innovation of Transformers is the attention mechanism, particularly the Scaled Dot-Product Attention. In this mechanism, each input is transformed into three vectors: Query (Q), Key (K), and Value (V). Q represents query vector that specifies the information a given token is seeking; the key vectors K represent the information that each token contains; and V represents the attention weight value that is computed using the alignment between query and key. The keys that are strongly aligned with the Query are weighted more, and if not aligned or are irrelevant then is they are assigned zero value.

The diagram provided in Figure 1.8 illustrates the high-level architecture of a Transformer model:

- *Input Sequence (Left Side)*: Tokens are converted into embeddings with added positional information and passed into the encoder.
- *Encoder (Blue Blocks)*: A stack of encoder layers processes the input, extracting contextual representations.
- *Attention Module (Center, Orange)*: It depicts the core attention mechanism with its three components: **Query (Q)**, **Key (K)**, and **Value (V)**. This

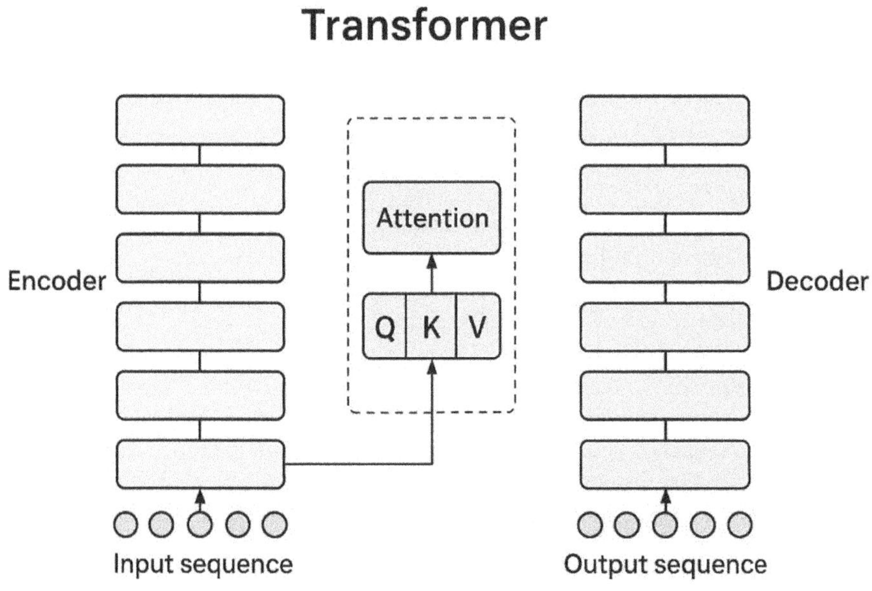

FIGURE 1.8 High-Level Architecture of a Transformer Model

block allows the model to weigh the importance of each token relative to others.

- *Decoder (Green Blocks)*: It takes the encoder output and the previously generated tokens to produce the final sequence.
- *Output Sequence (Right Side)*: It represents the final predictions (e.g., translated sentence, text generation output).

Arrows indicate the flow of information: from the input to the encoder, through the attention mechanism, and finally to the decoder for output generation.

1.9 LOSS FUNCTIONS IN GENAI MODELS

Loss function measures the learning capability of the model that generates the data with respect to the actual data with minimum errors. Various loss functions are discussed in the following sections.

1.9.1 Cross-Entropy Loss

Cross-entropy is the most common loss function, especially in NLP and text generation tasks. Cross-entropy computes the difference between the probability distribution predicted by the model and the true distribution (ground truth).

1.9.2 Kullback–Leibler (KL) Divergence

KL divergence quantifies the disparity between two probability distributions: the learned distribution and the actual data distribution. KL divergence is usually applied in VAEs to make the latent space obey a desired distribution (usually Gaussian).

The right choice of loss functions is very important as it affects generative quality and stability. The properly selected loss functions enhance fidelity, diversity, and coherence of the generated text. The inappropriately selected loss functions can cause unstable training or bad generalization.

1.10 METRICS FOR EVALUATING GENERATIVE MODELS

Generative models are hard to evaluate, as the outputs can be subjective and varied. Both qualitative and quantitative metrics are employed to measure their performance.

1.10.1 Perplexity

Perplexity is often used for language models to determine how well a model can predict a sequence of words. Less perplexity means that the model is better at predicting.

1.10.2 Bilingual Evaluation Understudy (BLEU)

BLEU compares n-grams of generated text with reference texts to measure text generation. It is commonly employed in machine translation and summarization.

1.10.3 Recall-Oriented Understudy for Gisting Evaluation (ROUGE)

ROUGE estimates overlap between generated and reference texts with emphasis on recall. It is frequently utilized for summarization, as well as assessment of text generation quality.

1.10.4 Fréchet Inception Distance (FID)

FID is applied to image-based generative models. FID compares the generated image distribution to actual images based on features obtained by a pretrained network (Inception-v3). Lower FID specifies greater realism of images.

1.10.5 Inception Score (IS)

An Inception Score assesses image quality and diversity by considering the confidence of class predictions in an Inception network.

1.10.6 Human Evaluation

Human judgment is still the best method to evaluate creativity, realism, and contextual appropriateness, particularly in open-ended activities where quantitative metrics are inadequate.

1.11 ETHICAL AND SOCIETAL CONSIDERATIONS

GenAI technologies provide groundbreaking opportunities across various fields, though they pose deep ethical and societal challenges. The adoption of GenAI technologies requires cautious consideration of risks of misuse, bias, fairness, accountability, and alignment with human values. The following section refers to three urgent issues: the emergence of deepfakes and disinformation, bias in generative models, and the significance of responsible AI and alignment.

1.11.1 Deepfakes and Misinformation

Generative models, particularly those that can generate hyper-realistic text, audio, and video, have made it possible to produce deepfakes—a real synthetic media that is indistinguishable from real content. Although deepfakes have positive applications in entertainment, education, and accessibility, their use for malicious purposes is dangerous. For instance, manipulated videos can be used to spread political disinformation, increase social unrest, or ruin reputations. The challenge is to balance innovation with protection, including strong detection capabilities, watermarking, and regulatory methods, to block abuse without hampering legitimate uses.

1.11.2 Bias in Generative Models

The AI systems are trained on large datasets. These training datasets may be biased on the basis of history, culture, or demographics. Such biases can show the output results such as stereotypes in a generated text, unfair representation in images, or biased recommendations. Bias not only undermines the fairness and credibility of AI but can also continue existing social inequalities when such systems are deployed at scale. Addressing bias requires diverse and representative training datasets, algorithmic interventions for fairness, and continuous auditing of outputs. Moreover, transparency in data sources and training processes is critical for fostering trust (Shuford, 2024).

1.11.3 Responsible AI and Alignment

The moral use of GenAI goes beyond bias removal to the general principle of Responsible AI. This encompasses responsibility for output, data privacy, and minimizing environmental impact of large-scale training. Key to overcoming the challenge is AI alignment—ensuring that generative models behave in a way that aligns with human intent, values, and societal norms. Misaligned systems can produce dangerous, deceptive, or offensive content without even malicious intent. Investigation in alignment involves reinforcement learning from human feedback (RLHF), interpretability methods, and governance framework development.

It makes the engagement of policymakers, industry representatives, and civil society in a collaborative effort to establish clear norms and regulations for the safe and productive use of GenAI. Policymakers can establish rules that shield individuals from harm, industry leaders can promote responsible technological development, and civil society organizations can advocate for public values and interests. By integrating these efforts, it is possible to develop frameworks that not only foster innovation but also ensure that fairness, transparency, and accountability in the application of GenAI are protected.

EXERCISES

MCQs

1. Which of the following is not considered a pillar of AI?
   a) ML
   b) DL
   c) GenAI
   d) Quantum computing
2. In Transformers, the attention mechanism is primarily used to:
   a) Reduce training loss
   b) Focus on relevant parts of input data
   c) Increase dataset size
   d) Improve GPU utilization
3. Which loss function measures the difference between two probability distributions?
   a) MSE
   b) Cross-entropy loss
   c) KL divergence
   d) Perplexity

4. The BLEU metric is mainly used to evaluate:
   a) Image quality
   b) Text generation against reference translations
   c) Model training efficiency
   d) Fairness of AI systems
5. Which of the following metrics is most commonly used for evaluating image generation models?
   a) BLEU
   b) ROUGE
   c) FID
   d) Perplexity
6. Deepfakes are primarily associated with:
   a) Data augmentation
   b) Synthetic media manipulation
   c) Model optimization
   d) Loss function tuning
7. Responsible AI aims to ensure:
   a) Faster training
   b) Alignment with ethical and societal values
   c) Higher computational efficiency
   d) Better GPU usage

Short/Long Answer Questions

1. Define GenAI and state why it is different from classical AI.
2. What is the role of fairness in GenAI?
3. Enumerate the key pillars of AI presented in this chapter.
4. What is the purpose of the attention mechanism in transformers?
5. Identify two loss functions that are frequently utilized in GenAI.
6. Distinguish between BLEU and ROUGE metrics.
7. Define deepfakes and state why they are deemed to be harmful.
8. Provide an example of bias in generative models.
9. Discuss the historical development of GenAI and key milestones that shaped its progress.
10. Explain the AI pillars—ML, DL, GenAI, and LLMs—highlighting their interrelationships.
11. Describe the GenAI blueprint, covering training data, foundation model training, and the generation phase.
12. How do Transformers and the attention mechanism revolutionize generative models?
13. Compare and contrast cross-entropy loss and KL divergence with suitable examples.
14. Evaluate different metrics (BLEU, ROUGE, FID, IS, human evaluation) for assessing generative models, discussing strengths and limitations.
15. Critically analyze the ethical challenges posed by generative models, including misinformation, bias, and alignment.

Answers to MCQs

1. d) Quantum computing
2. b) Focus on relevant parts of input data
3. c) KL divergence
4. b) Text generation against reference translations
5. c) FID
6. b) Synthetic media manipulation
7. b) Alignment with ethical and societal values

REFERENCES

Bengesi, S., El-Sayed, H., Sarker, M. K., Houkpati, Y., Irungu, J., & Oladunni, T. (2024). Advancements in generative AI: A comprehensive review of GANs, GPT, autoencoders, diffusion model, and transformers. *IEEe Access*, *12*, 69812–69837.

Blattmann, A., Dockhorn, T., Kulal, S., Mendelevitch, D., Kilian, M., Lorenz, D., . . . & Rombach, R. (2023). Stable video diffusion: Scaling latent video diffusion models to large datasets. *arXiv preprint arXiv:2311.15127.*

Bory, P. (2019). Deep new: The shifting narratives of artificial intelligence from Deep Blue to AlphaGo. *Convergence*, *25*(4), 627–642.

Chang, Y., Wang, X., Wang, J., Wu, Y., Yang, L., Zhu, K., . . . & Xie, X. (2024). A survey on evaluation of large language models. *ACM Transactions on Intelligent Systems and Technology*, *15*(3), 1–45.

Choudhry, M. D., Sundarrajan, M., & Sundaram, K. (2024). Bias and fairness in generative AI. *Generative AI and LLMs: Natural Language Processing and Generative Adversarial Networks*, 177.

Dally, W. J., Keckler, S. W., & Kirk, D. B. (2021). Evolution of the graphics processing unit (GPU). *IEEE Micro*, *41*(6), 42–51.

Estebon, M. D. (1997). Perceptrons: An associative learning network. *Virginia Tech.—1997.*

Goodfellow, I., Pouget-Abadie, J., Mirza, M., Xu, B., Warde-Farley, D., Ozair, S., . . . & Bengio, Y. (2014). Generative adversarial nets. *Advances in Neural Information Processing Systems*, 27.

He, R., Cao, J., & Tan, T. (2025). Generative artificial intelligence: A historical perspective. *National Science Review*, *12*(5), nwaf050.

Hinton, G. E., Osindero, S., & Teh, Y. W. (2006). A fast learning algorithm for deep belief nets. *Neural Computation*, *18*(7), 1527–1554.

Hu, Y. J., Yeh, C. L., & Laun, W. (2009, November). Challenges for rule systems on the web. In *International Workshop on Rules and Rule Markup Languages for the Semantic Web* (pp. 4–16). Berlin, Heidelberg: Springer.

Petrov, M., & Wortmann, T. (2021). Latent fitness landscapes-exploring performance within the latent space of post-optimization results. *Proceedings of the SimAud, Online*, 15–17.

Rhee, E. J. (2018). A deep learning approach for classification of cloud image patches on small datasets. *Journal of Information & Communication Convergence Engineering*, *16*(3), 173–178.

Saon, G., & Picheny, M. (2017). Recent advances in conversational speech recognition using convolutional and recurrent neural networks. *IBM Journal of Research and Development*, *61*(4/5), 1–1.

Schneider, J. (2024). Explainable Generative AI (GenXAI): A survey, conceptualization, and research agenda. *Artificial Intelligence Review*, *57*(11), 289.

Sharma, A. K., & Sharma, R. (2024). Generative Artificial Intelligence and legal frameworks: Identifying challenges and proposing regulatory reforms. *Kutafin Law Review, 11*(3), 415–451.

Shuford, J. (2024). Examining ethical aspects of AI: Addressing bias and equity in the discipline. *Journal of Artificial Intelligence General science (JAIGS) ISSN: 3006–4023, 3*(1), 262–280.

Vaswani, A., Shazeer, N., Parmar, N., Uszkoreit, J., Jones, L., Gomez, A. N., . . . & Polosukhin, I. (2017). Attention is all you need. *Advances in Neural Information Processing Systems, 30.*

Wang, H., Lei, Z., Zhang, X., Zhou, B., & Peng, J. (2016). Machine learning basics. *Deep Learning*, 98–164.

Xiang, A. (2024). Fairness & privacy in an age of generative ai. *Science and Technology Law Review, 25*(2).

Zong, M., & Krishnamachari, B. (2022). A survey on GPT-3. *arXiv preprint arXiv:2212.00857.*

2 The Ethical Landscape of GenAI

Over the last decade, the world of computer science has shifted significantly because of GenAI. This new type of AI may mean that we treat machines less as tools and more as creative partners. While early AI mostly performed classification, prediction, and simple optimization, GenAI can write stories, paint pictures, compose music, and even propose molecular designs. The rise of models such as GANs, Transformer-based language systems, and newer diffusion models shows how fast things have moved (Vaswani et al., 2017). Researchers such as Goodfellow and Vaswani introduced ideas that appear to copy some aspects of human thought; however, they still lack true understanding (Goodfellow et al., 2014).

What powered this surge? On the technical side, huge data collections and faster chips double their speed very often, and new tricks in DL push capabilities higher (Marcus & Davis, 2020). From a social perspective, people want more automation, personalized experiences, and fresh ideas. Therefore, GenAI feels like a game-changer in health, banking, education, art, and even government. However, there are concerns. Some argue that relying on generated content can erode critical thinking. Others have pointed out that biases hidden in the training data may spread more widely. Thus, while GenAI reshapes daily tasks, choices, and even the movies we watch, it also forces us to ask if we are handing away too much creative control today (Crawford, 2021).

With all these advances, the ethics of GenAI seem more urgent now. Many scholars and workers fear that these tools could copy or even make worse existing social gaps. They worry it could shake trust in news info spaces, and test laws, cultures, and moral rules we have. While narrow AI mostly sticks to clear, limited jobs, GenAI brings new moral puzzles because it can act creatively and on its own (Kirchner, 2016). For example, a generative system might write a press article, a medical note, or a legal filing that looks real but is totally fake. Those fake items could be hard to tell from real ones. This mix of fact and fiction still may appear to threaten democratic politics, public trust, and personal freedom (Chesney & Citron, 2019).

Additionally, GenAI's ethical concerns extend beyond misinformation. Essentially, generative models work with large sets of data obtained from the internet and private sets. These datasets are often embedded with historical biases, stereotypes, and cultural differences that are later present in outputs generated by these models (Barocas, Hardt, & Narayanan, 2019). For instance, when deployed during hiring procedures, a generative system can reinforce trends of discrimination and thus de facto disqualify minority groups. Similarly, in creative fields, authors have expressed concerns about being used to train AI models without permission to create their work, thus questioning the basis of intellectual property laws (Samuelson, 2023). Thus, the GenAI's moral environment is shaped by

DOI: 10.1201/9781003713951-2

double tension: GenAI holds out hopes to enhance human imagination, expedite scientific discovery, and enhance societal welfare. It can deepen inequality, facilitate its misuse, and undermine basic human rights. It is imperative to address this tension by employing multidisciplinary efforts that cross computer science and engineering disciplines.

There is a requirement for philosophers and ethicists to work with policymakers and legal experts to join efforts with technologists to create governance frameworks to achieve a balance between innovation and accountability (Jobin, Ienca, & Vayena, 2019). Institutions such as the United Nations Educational, Scientific, and Cultural Organization (UNESCO), Organisation for Economic Co-operation and Development (OECD), and the European Union (EU) have created guidelines and regulatory frameworks to ensure that AI development remains aligned with human values (European Commission, 2021). The rate at which technology is growing is significantly greater than policy implementation rates, with significant gaps in oversight and accountability. This chapter aims to chart the ethical landscape of GenAI by exploring its benefits and risks. Section 2.2 outlines the prospects of GenAI in science, art, and industry. Section 2.3 examines the risks and side effects involved in its application. Section 2.4 takes a closer look at the root ethical issues—bias, fairness, transparency; accountability, intellectual property, and privacy—while Section 2.5 takes a critical look at the adoption benefits and drawbacks through key sectors.

Section 2.6 examines changing regulations and ethical landscapes, and Section 2.7 suggests directions for the responsible and human-centric application of GenAI. Finally, this chapter concludes by considering how societies can reap the most benefits from GenAI while reducing its ills. Essentially, GenAI is as much a technology as it is a social phenomenon. Its direction will determine the future trajectory of knowledge, imagination, and governance.

Whether it will become a vector for benevolent progress or an instrument for causing harm is contingent upon the ethical decisions being made today. Therefore, examining ethical landscape of GenAI is an academic pursuit and an imperative for humanity's shared future. Figure 2.1 shows the promises of GenAI.

2.1 THE PROMISE OF GENAI

GenAI started out as a weird lab toy. Now it feels like a switch that flips whole fields on fast. It is not just spotting patterns like old programs did. It actually makes new stuff—sentences, pictures, movies, even chemical formulas and engineering plans. One analyst said it "creates new insights, new representations." That sounds like a big leap, a real paradigm shift, even if the wording sounds fancy. The idea is that AI can push human knowledge farther, give more people tools they never had, and maybe ease some big world problems.

Think about science—the models can guess things that we had not seen before, give us fresh ideas, new art pieces, faster robots, different ways to teach, and even greener ways to live. All of those use the same trick: the AI looks past the data we fed it and writes something brand new. In this section, we outline the prospects for GenAI in some key areas: scientific research, culture and imagination, productivity

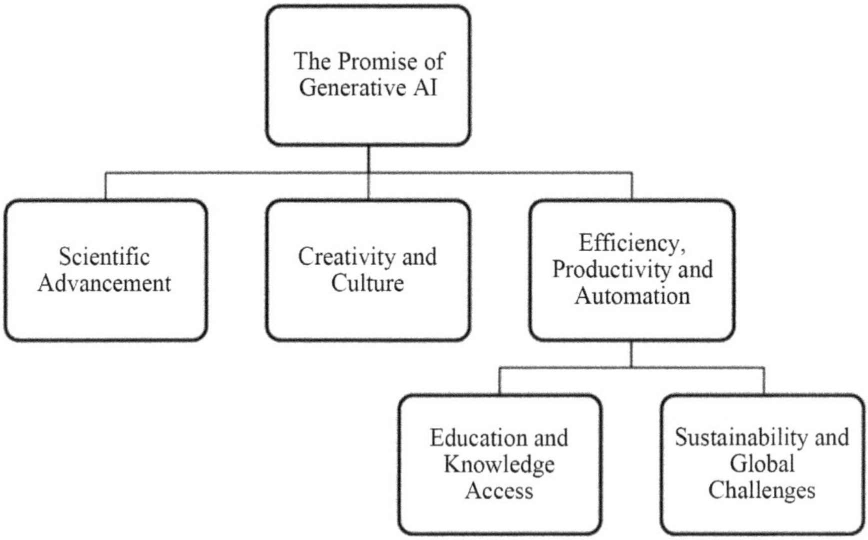

FIGURE 2.1 Promise and Pitfalls for GenAI

and robotics, education, and sustainability. These prospects are based on GenAI's potential to deepen human knowledge, equalize access to resources, and overcome some of society's biggest challenges.

2.1.1 Scientific Advancement

One of GenAI's most powerful commitments is scientific discovery. Much scientific research involves tedious experimentation, simulations, or data-gathering procedures. However, GenAI can simulate, create, and foretell difficult phenomena at a fraction of the innovation cycle previously possible. A prime example is prediction of protein structures.

For decades, three-dimensional (3D) protein structure prediction from amino acid sequences has been a grand challenge in molecular biology. In 2021, DeepMind's AlphaFold AI system predicted protein structures with experimental-level accuracy using generative modeling (Jumper et al., 2021).

The publication of more than 200 million predicted protein structures through the AlphaFold Protein Structure Database (AlphaFold DB) was a paradigm-breaking breakthrough for biology and presented scientists with an unprecedented tool for drug discovery, gene and heredity research, and disease therapy. Similarly, in drug discovery, generative models are used to design novel molecules with specific chemical properties (Helbing, 2019). Traditional drug development can take over a decade and cost billions of dollars.

GenAI-enabled approaches can generate candidate molecules, simulate interactions with biological targets, and optimize their chemical properties in a fraction of the time required by conventional methods (Zhavoronkov et al., 2019). During the COVID-19 pandemic, generative algorithms contributed to the rapid exploration of antiviral compounds, underscoring their importance in public health emergencies.

Generative models are also gaining traction in the fields of physics and engineering. Materials science applies GenAI to suggest new alloys and nanomaterials with optimal mechanical or thermal attributes (Xie et al., 2021). Aerospace engineers seem to use generative design for building parts that feel both light and strong, reducing material waste. This shift may show how GenAI is stretching what humans can study, reaching once-unreachable scientific ground in fields such as space travel.

2.1.2 Creativity and Culture

Outside the lab, GenAI seems to spark the imagination. DALL-E, Stable Diffusion, and ChatGPT allow anyone to try art once limited to experts. Artists, writers, and musicians mix their ideas with algorithmic power. A gallery shows a diffusion picture, prompting discussions about real creativity and ownership.

In 2018, Christie's sold "Edmond de Belamy," a portrait created by a GAN. The piece fetched $432,500, which raises doubts: does a computer-made image deserve such value, or is the market just chasing novelty in the art world? That sale may show both money value and cultural weight of AI-made art (Elgammal et al., 2017). For music, OpenAI's Jukebox lets creators blend genres or even imitate a baroque composer, opening paths that were previously rare.

GenAI also appears to help keep culture alive. AI has been used to repair faded paintings, rebuild lost historic places, and even shape reconstructions of extinct languages and scripts (Kenderdine, 2020). By adding to collective memory, it could make heritage reachable for kids in futures we cannot yet picture. Still, questions about originality and ownership remain hanging in the air. Yet the main merit of GenAI might be that it gives people who never called themselves artists a chance to express themselves. Therefore, culture becomes less a club for the trained few and more a playground for many.

Teachers use these tools in class, hoping they spark curiosity but worry they might replace. By augmenting cultural memory, GenAI functions as a means of preservation and inclusivity, guaranteeing that human heritage is attainable for future generations. Even while questions of originality and ownership remain unresolved, GenAI's merit is that it allows those who are normally artistic novices to access creative expression, thereby democratizing access to culture.

2.1.3 Efficiency, Productivity, and Automation

GenAI is equally revolutionary in boosting productivity and automating various sectors. Knowledge work, formerly believed to resist automation, is being reengineered by GenAI-driven assistants. For instance, language models such as GPT-4 and Claude are capable of writing reports, condensing legal agreements, designing computer codes, and even developing marketing campaigns.

In software development, GenAI tools, such as GitHub Copilot, suggest code snippets, automate debugging, and improve developer productivity (Vaithilingam et al., 2022). This not only accelerates software production but also lowers the entry barrier for novice programmers, thereby creating a more inclusive technological ecosystem. In business operations, generative models streamline customer support by powering Chabot that provides 24/7 multilingual assistance. They also aid document generation, financial reporting, and predictive analytics. These applications reduce repetitive workloads, allowing humans to focus on higher-order problem solving and innovation.

Furthermore, GenAI has potential applications in journalism and media. Automated content generation tools can produce financial summaries, weather reports, and sports recaps, freeing journalists to pursue investigative reporting. Although ethical concerns persist regarding misinformation (to be discussed in Section 2.3), the productivity gains of GenAI in content-heavy industries are undeniable.

2.1.3.1 Education and Knowledge Access

GenAI could change the way we learn, but the real impact is uncertain. It might let virtual tutors match each student's speed, style, and need, something hard to get in a big classroom (Luckin et al., 2016). For example, a model can make personal practice quizzes, hold short chat dialogs, or rewrite tough ideas in plain language.

In addition, this tech probably widens access to education. AI can write summaries, translate text, or turn speech into written words, which could lower barriers for disabled learners or people in language-poor regions. Automatic translators powered by AI are already trying to close English-centered knowledge gap in developing countries.

At university level, GenAI appears to help with searching papers, forming hypotheses, and making sense of data, cutting down hours spent on boring chores. If schools adopt these tools, they may give students more room for critical thinking, letting them focus on synthesis and new ideas. In conclusion, the promise is big, yet we still need very careful testing.

2.1.3.2 Sustainability and Global Challenges

GenAI might help solve big world sustainability problems. It seems to improve power grids, show climate trends, and make greener materials. One example: GenAI builds models that picture renewable-energy options, so governments and firms could plan slower transitions (Rolnick et al., 2022). In addition, these tools can guess crop yields, water use, and pest plans, which could boost food safety.

In ecology, GenAI appears to map biodiversity, pointing out how ecosystems may shift under warming. Yet, results may be uncertain, and the tech needs careful oversight. Many think these methods support the UN's Sustainable Development Goals, although they are not a magic fix or at least a step forward.

2.1.3.3 Case Study: GenAI in COVID-19 Response

The COVID-19 outbreak gave a real-world look at how GenAI could help during a crisis. Researchers used generative models to copy the protein shape of

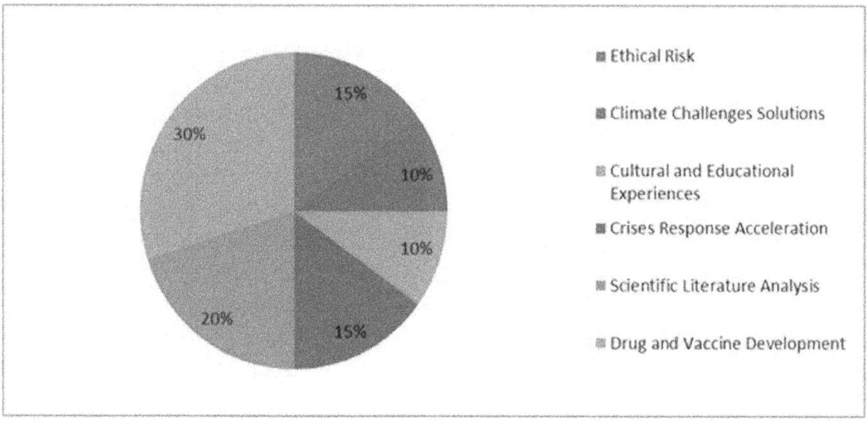

FIGURE 2.2 GenAI in Crisis Management and Beyond

SARS-CoV-2, to suggest antiviral molecules, and even to improve vaccine designs (Chenthamarakshan et al., 2020). At the same time, AI-driven text tools scanned every new paper on platforms like CORD-19, making the surge of data easier to digest. Those uses seem to show that GenAI can act like a force multiplier, cutting down response time, and urging quick scientific teamwork.

The promise of GenAI is not merely to copy what we already do. This may mean faster drug research, new ways to make videos, and easier access to courses, even tools that help track carbon emissions. These chances feel almost like a turning point for me. However, these benefits are not just for scientists; a school student could use the program to design a science fair project, which might close the gap in ideas (Boddington, 2017). However, this power raises significant ethical concerns. Misinformation could spread, bias might slip in, and the technology could be used for bad purposes. Therefore, each hopeful gain must be measured against these risks. The debate is ongoing in many countries. Figure 2.2 shows summary about crisis management by GenAI.

2.1.4 Ethical Challenges and Implications

GenAI appears to have turned the online world upside down by allowing computers to create stories or pictures that seem to be made by people. However, this power brings a bundle of ethical puzzles that we must face. Unlike rule-driven software, GenAI works on chance, hides its own steps, and pulls from large datasets. Therefore, outcomes may be tougher to guess, choices harder to read, and blame tougher to place. Some consider it helpful, while others fear cheating (Burrell, 2016; Mittelstadt et al., 2016). The big worries cover bias and unfair treatment, fake news and deepfakes, who owns the work, privacy slips, jobs being pushed out, lack clear accountability, environmental impact, and still weak oversight. Each point feels more a problem than a tech glitch, shaping democracy, rights, and the planet's fairness.

2.2 BIAS AND DISCRIMINATION

GenAI systems are built using vast amounts of internet data, which often reflect old biases and stereotypes. Therefore, these prejudices seem to crawl into the model's answers, possibly worsening discrimination. Research shows text-to-image tools, such as Stable Diffusion, still link "nurse" mostly with women and "CEO" with men (Cho et al., 2023). Therefore, the outputs may reinforce gender gaps, although some developers argue that they can be fixed later in future versions of the software.

- *Bias in language*: GPT-4 and other LLMs can create content favoring English-speaking, Western cultures' knowledge over neglected cultures and dialects (Bender et al., 2021). Bias in GenAI is a serious concern in high-stakes usage, such as employment, lending, and medicine, where biased outputs can perpetuate inequalities and engender algorithmic unfairness (Noble, 2018).

Impact: Fairness-sensitive training, bias auditing, and diversified datasets are required for developers and officials. Even perfect neutrality may become unavailable, and one of the resulting ethical issues is the values incorporated in AI systems (Bryson, 2018).

2.2.1 Misinformation and Deepfakes

The GenAI lowered the barrier to creating realistic but fictional content, and there is apprehension about undermining trust in public information networks.

- *Political Deepfakes*: Political leaders have already been spoofed using synthesized videos with the potential for election interference and misinformation campaigns (Chesney & Citron, 2019).
- *Generative Text*: Social media robots and machine-generated newspapers can propagate misinformation on a large scale, overloading systems for checking facts (Zellers et al., 2019).

Impact: The spread of artificial content is an affront to democratic institutions, journalism, and civic conversations. Ethical countermeasures include watermarking, provenance tracking, and civic education programs.

2.2.2 Intellectual Property and Authorship

The most contentious GenAI-based ethical argument is probably who owns the work created by AI and whether training on copyrighted materials is an infringement.

- *Copyright violation*: Certain instances have been filed against AI companies for using copyrighted images, songs, and literature without permission (Samuelson, 2023).
- *Ambiguity regarding authorship*: Is AI an author, the developer, a user, or a provider of data? Who authored it and whose work was used in training? Existing IP laws are ambiguous on this (Gervais, 2020).

Implications: Innovation must not come at the expense of creators' rights. Without principles, the GenAI will tear down the creative economy and discourage original human writing (Cath, 2018).

2.2.3 Privacy and Data Protection

GenAI systems are usually trained with personal data scraped from the open internet, which may cause inadvertent reprinting of sensitive data.

- *Case Study*: The researchers confirmed that LLMs might occasionally "leak" personal information such as phone numbers, email addresses, or even health records that are part of training sets (Carlini et al., 2021).
- *Synthetic Data Paradox*: Even though GenAI can crank out "anonymous" fake data sets, attackers sometimes manage to trace the original people behind them (Yoon et al., 2020).

Implications: That could mean privacy rules like General Data Protection Regulation (GDPR) or Health Insurance Portability and Accountability Act (HIPAA) get shaky. Therefore, companies should probably stick to data minimization, clear consent, and learning methods that care about privacy (Dignum, 2019).

2.2.4 Labor Displacement and Economic Inequality

At the same time, GenAI seems to reshape the job market. It takes over tasks that used to need human journalists, graphic designers, translators, or coders. Newsrooms use AI for routine stories, which may potentially lessen need for reporters (Marconi & Siegman, 2017). Yet some fear creativity could suffer.

- *Creative Industries*: Designers and artists report revenue losses due to AI software offering cheap replacements (Elkins, 2023). Although AI offers gains in productivity, the sharing of economic gains is uneven and stirs up technological unemployment and greater inequality.

Import: Ethical adoption requires reskilling policies, fair remuneration, and social protection nets to shield vulnerable workers.

2.2.5 Transparency and Accountability

The black-box property of generative models makes it difficult to explain the output generation.

- *Black-box Problem*: There are billions of parameters in Transformer-based models such as GPT-4, and their internal reasoning is almost never interpretable (Lipton, 2018).

- *Accountability Gap*: There is ambiguity regarding who is responsible (developer, deployer, or user) if harmful outputs are generated (e.g., defamatory writing or discriminatory hiring decisions).

Impact: There is a growing demand for explainable AI (XAI) and algorithmic impact statements, but practical deployment is not in sight (Eubank, 2018).

2.2.6 Environmental Impact

Training large generative models is an energy-intensive and power-hungry process.

- *Carbon Footprint*: The training of one large NLP model was estimated by Strubell et al. (2019) to release an amount of carbon comparable to five cars throughout their lifetimes.
- *Resource Allocation Inequities*: The huge expense of training the model concentrates power in giant tech corporations, leaving less room for small players and new economies (Crawford, 2021).

Implications: The development of AI requires green AI strategies, energy-efficient designs, and carbon responsibilities (Feldstien 2019).

2.2.7 Governance and Global Regulation

The regulation of GenAI is still dispersed, as national policies, trade self-regulation, and international guidelines remain fluid and inconsistent (Gillespie, 2018).

- *Regional Variations*: The EU AI Act favors risk-based regulation, whereas the United States prefers sector-specific guidelines. China imposes rigid controls over deepfakes (Creemers, 2022, 2023).
- *Global Challenge*: Without universal harmonization, uneven rules threaten the formation of AI havens where ethical regulation is the lowest.

Implications: A balance between protection and innovation must be struck, and international norms concerning safety, fairness, and human rights must be encouraged (Heaven, 2020).

2.2.8 Ethical Trade-Offs

Deciding where to draw lines feels tougher when values clash.

- *Transparency against intellectual property*: how much training data *might* be revealed?
- *Privacy versus innovation*: can we still feed data to GenAI without possibly invading privacy?
- *Freedom of expression versus harm reduction*: should we block content?

These choices rarely have clear answers. These paradoxes suggest that good governance may not mean eliminating every risk. Instead, it could be about balancing trade-offs. The ethical questions surrounding GenAI do not appear to be mere technical problems (Hern, 2019). They feel like larger social issues that spill into culture, law, and the economy. Who benefits, who is left out, and who is held answerable?

Solving these problems probably requires more than technical guards. New moral ideas, broader policies, and global teamwork are required (Huang, 2020). The worry is not about total control. It is about finding ways to steer AI so that dignity, fairness, and justice remain alive today. Schools must also teach responsible use (Kerr, 2019).

Ethical Challenge	Examples/Evidence	Implications
Bias and Discrimination	• Text-to-image models reinforce gender stereotypes Cho et al. (2023). • LLMs privilege Western-centric knowledge Bender et al. (2021).	Risk of algorithmic injustice in hiring, lending, and healthcare. Requires fairness-aware training, audits, and diverse datasets—though perfect neutrality is impossible.
Misinformation and Deepfakes	• Synthetic videos impersonating leaders Chesney and Citron (2019). • AI-generated news spreading false info Zellers et al. (2019).	Undermines democracy, journalism, and trust. Solutions include watermarking, provenance tracking, and media literacy.
Intellectual Property and Authorship	• Lawsuits over training on copyrighted works Samuelson (2023). • Ambiguity of authorship Gervais (2020).	Unclear ownership may undermine creative economy. Need legal frameworks to balance innovation and creators' rights.
Privacy and Data Protection	• LLMs "leak" private info Carlini et al. (2021). • Synthetic data vulnerable to re-identification Yoon et al. (2020).	Compliance with GDPR/HIPAA essential. Requires data minimization, consent, and privacy-preserving learning.
Labor Displacement and Inequality	• AI-generated news articles Marconi and Siegman (2017). • Artists losing income to AI tools Elkins (2023).	Risk of job loss and inequality. Policies needed for reskilling, fair pay, and safety nets.
Transparency and Accountability	• Black-box models like GPT-4 hard to interpret Lipton (2018). • Accountability gap for harmful outputs.	Push for XAI and impact assessments, though implementation remains difficult.
Environmental Impact	• Training large NLP models = carbon of five cars Strubell et al. (2019). • High costs centralize power Crawford (2021).	Need for green AI, efficient architectures, and carbon accountability.
Governance and Regulation	• EU AI Act versus US guidelines versus China's strict controls EU (2023) and Creemers (2022).	Risk of "AI havens" without global coordination. Must balance innovation with human rights and fairness.
Ethical Trade-offs	• Transparency versus IP • Privacy versus Innovation • Expression versus Harm reduction	Ethical governance is about managing trade-offs responsibly, not eliminating risks.

2.3 PROS AND CONS OF AI ADOPTION ACROSS INDUSTRIES

Technology brings fast automation, a bit of personal feel, and ideas that seemed impossible not long ago. However, the spread of these tools is not all good; clear dangers may appear. This may increase efficiency, open access, and even spark inventions (Krafft et al., 2020). Simultaneously, it could make us dependent, widen gaps, and leave systems exposed. Hospitals enjoy quicker diagnostics, and banks use smarter apps. Schools try to create customized lessons.

Governments are testing AI policies. Courts are concerned about algorithmic bias. Newsrooms try auto-written stories. Artists play with new tools, such as AI. Researchers have accelerated the experiments (IBM, 2020).

2.3.1 Healthcare

Pros

- *Better Imaging and Diagnostics*: Generative models may produce magnetic resonance imaging (MRI) or computed tomography (CT) scans when data are lacking. This could improve cancer, brain, or heart diagnosis while synthetic data keep patient privacy, though risks remain (Yi et al., 2019).
- *Drug Discovery and Personalized Medicine*: AI may predict molecular structures and repurpose drugs, likely slashing research time and costs (Zhavoronkov et al., 2019). Yet predictions are not always reliable, and safety checks stay needed.
- *Medical Education and Training*: Simulated virtual patients give trainees practice in a safe setting. They engage learners, though some feel scenarios lack the messiness of real clinics.

Cons

- Data privacy remains a concern, even with synthetic data. A badly masked source reveals a patient's name or age, according to Kaissis et al. (2020). Moreover, many algorithms do not contain enough minority examples, which may cause biased results and widen gaps (Obermeyer et al., 2019). Therefore, researchers should review privacy protections and dataset diversity in their work.
- Relying too heavily on AI for diagnosis raises questions regarding accountability. Who is liable: the AI vendor, the hospital, or the clinician? This uncertainty suggests a potential need for a clearer responsibility framework (Leslie, 2019).

2.3.2 Finance

Pros

- Fraud detection may benefit from generator models that copy shady behavioral patterns, possibly boosting anomaly detection (King et al., 2020).

- Personalized financial help appears in AI chatbots and virtual advisers, offering users tailored guidance on money matters.
- Market forecasting could use GenAI to sketch future scenarios, perhaps shaping investment choices for better outcomes (Lin et al., 2017).

Cons

- *Bias in Credit Scoring*: If there is discrimination in historic lending data, GenAI models can reinforce financial exclusion (Hurley & Adebayo, 2017).
- *Synthetic Market Manipulation*: Computer-generated false news or trade signals can disrupt markets.
- *Cybersecurity Risks*: Generative models can be exploited to generate phishing email headers or fake identities for fraud.

2.3.3 Education

Pros

- *Personalized Learning Pathways*: GenAI-powered tutors generate individualized content, thereby adapting to each student's pace and style (Luckin, 2018).
- *Accessibility and Inclusion*: Students with disabilities benefit from AI-generated transcripts, summaries, and translations.
- *Content Generation*: Teachers save time by automating quizzes, assignments, and lesson plans (Lu, 2019).

Cons

- *Academic Integrity*: Students may misuse text generators for plagiarism, undermining learning outcomes (Cotton et al., 2023).
- *Equity Gap*: Elite institutions may afford advanced AI tutors, while underfunded schools risk being left behind.
- *Depersonalization of Teaching*: Excessive reliance on AI may erode critical thinking and human mentorship.

2.3.4 Governance and Law

The meeting point of GenAI and government rules looks like a big change, but it also brings a lot of doubts (Mittelstadt, 2019). Officials are starting to try AI tools to help them decide what to do, to cut down paperwork, and to talk more with citizens. Yet the idea opens questions about who is really in control, whether the outcomes stay fair, and whether the system ends up watching everyone too closely.

2.3.4.1 Advantages of GenAI in Governance and Law

GenAI seems to bring changes to government and law. It can make work faster and easier by running policy simulations and predicting results (Moor, 2006). Governments can attempt to predict the socioeconomic effects before they occur.

It can also talk to citizens with auto-messages, which might make things clear and quick.

In courts, AI helps write contracts, summarize case law, and conduct research, so lawyers might have less to do and spend less money (Floridi, 2019). Chatbots for legal aid may reach people who receive little help. Overall, AI can make systems smoother, more accurate, and truly open. Figure 2.3 gives an idea about Governance and Law for GenAI:

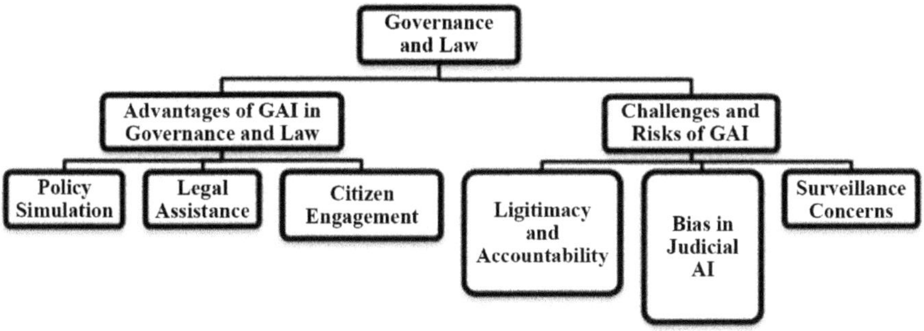

FIGURE 2.3 Governance and Laws for GenAI

2.3.4.1.1 Policy Simulation

When it comes to hard problems—climate shift, money gaps, health scares—AI may mean they can run detailed simulations before a law is even written. A ministry could, for example, model what a carbon tax might do to emissions over ten years, or picture how different vaccine plans could slow a flu wave (Hao, 2020). By showing many possible futures, the technology could point out hidden trade-offs and new chances, which might help rulers avoid a trial-and-error style. Some scholars even argue this could make policy more evidence based, though critics say models can hide biases.

By establishing a variety of future scenarios, GenAI helps policymakers discern trade-offs, surprises, and best practices (Brynjolfsson & McAfee, 2017). Such tools strengthen evidence-based policymaking and reduce reliance on trial-and-error techniques among public policies.

2.3.4.1.2 Legal Assistance

On the legal side, AI has started to take over many time-heavy tasks. It can draft simple contracts, pull short summaries from long court opinions, and even guess how a judge may rule based on past decisions. Such help could lift the load on lawyers and give people who cannot afford a full lawyer a chance to get usable papers or cheap advice (Surden, 2019). Small firms can generate binding agreements at a fraction of the cost (Remus & Levy, 2016).

However, the picture is not entirely bright. Automating law might spread errors quickly, and there is fear that the lack of human judgment could damage trust. Balancing efficiency and fairness will be key as societies move forward.

2.3.4.1.3 Citizen Engagement

Governments feel pressure to give quicker, clearer, fairer services. A demand from citizens seems to push this need. Chatbots and virtual assistants, built on GenAI, may help. These tools stay online all day, answering questions. People may ask about taxes, hospitals, or IDs (Wirtz et al., 2019).

The system replies fast often and feels personal. When government staff are low, such bots could ease long waits. That may mean more trust between citizens and officials. More people get help if they lack internet skills. Thus, the government appears more inclusive and open. Overall, this could encourage a participatory form of rule.

2.3.4.2 Challenges and Risks of GenAI in Governance and Law

GenAI might change how governments and courts work, but it also raises concerns. The legitimacy of decisions made by AI may be questioned, especially when laws are involved (Hagendorff, 2020). Bias in the data could lead to unfair judgments, which could break public trust in AI. Excessive reliance on AI may reduce human oversight, raising ethical concerns. Surveillance features in AI could threaten privacy if officials misuse them. Some argue that strict rules could limit innovation, yet without open frameworks, the technology could widen inequality. In short, without clear safeguards, democratic governance can still suffer.

2.3.4.2.1 Legitimacy and Accountability

Putting GenAI into government may shake ideas of who is legit and who is answerable. The way the machines pick choices is often hidden and built on mathematics most people and lawmakers do not get. Citron (2020) says handing over big decisions to AI could hurt openness, because citizens might not know whom to blame when stuff goes wrong. Imagine an AI that says no to welfare; people would not see how to fight that decision or who to blame (Zuboff et al., 2019). This foggy feel can chip away at trust in the system and make AI rule seem shaky for citizens.

2.3.4.2.2 Bias in Judicial AI

The use of GenAI in government may go against what we think of as true democracy. The idea that computers decide big stuff is not always clear. Citron (2020) states that this could hurt transparency because people might not know whom to blame when something goes wrong.

Consider a welfare program that uses AI to reject a claim. The applicant might not know the steps to fight the decision or who is responsible (Zeng et al., 2019). This uncertainty can shake the trust of the entire institution. It also makes us wonder whether AI rule-making is legitimate. Therefore, questions about accountability remain unanswered in future policy debates across societies worldwide and beyond (Winfield & Jirotka, 2018).

2.3.4.2.3 *Surveillance Concerns*

The use of GenAI in governance is often mixed with surveillance tools and raises privacy and civil rights questions. Governments may use AI to watch citizens' messages, movements, and even travel, claiming it is for safety. When this is added to facial recognition or prediction-based policing, GenAI could allow real-time, large-scale surveillance (Feldstein, 2019). Some benefits may be observed for police work, such as faster checks. However, the danger is that permanent mass watching could become normal, that people may feel afraid to speak, and that authoritarian rule could strengthen. To maintain freedom, strong laws and clear ethical rules are likely required to prevent GenAI from becoming a significant spying gadget (Whittlestone et al., 2019).

GenAI also provides big opportunities for government and law, but it brings tough problems too. This may help design policies, provide more people with a way to obtain legal help, and boost citizen involvement (Rahwan, 2018). However, it could weaken trust in justice, introduce bias into decisions, and lead to sloppy surveillance. The future of GenAI will probably rest on finding a middle ground between new technology and strong accountability (Russell et al., 2015).

Policymakers should write ethical guidelines, keep a watchful eye on regulations, and stay open so that using GenAI adds to democracy rather than hurting the rule of law. Public debate is also needed regarding where the line should be.

2.3.5 Media and Communication

GenAI appears to be reshaping media production. It reduces the cost of creating content, allowing small creators to publish more. It also allows people to communicate across languages, widening their voices. Real-time updates appear to be faster than ever. However, the same power causes problems. Fake news can travel instantly, threatening the truth. Journalists may lose some professional standards, and readers can feel overwhelmed by too much information. Thus, the technology brings both help and risk to public discourse. Figure 2.4 shows a flowchart about details of GenAI in media and communication.

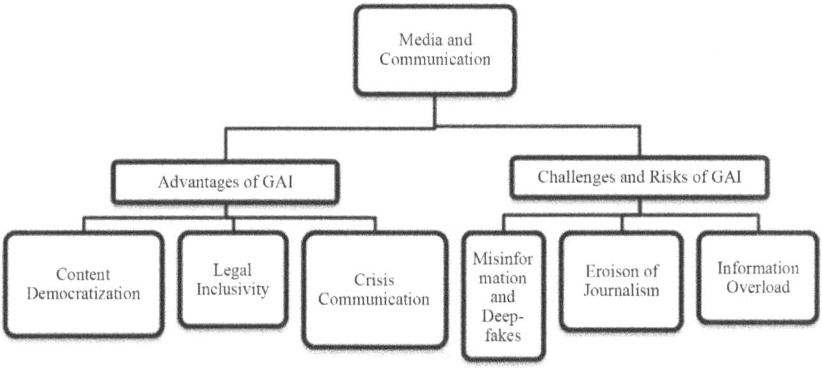

FIGURE 2.4 Media and Communication

2.3.5.1 Advantages of GenAI in Media and Communication

GenAI may change media and how we chat, shaping what we show, how it is built, and who sees it. It can spit out text, pictures, sounds, and videos, which can save time and money while sparking ideas. Journalists might use it for short news bits about a school play, for the translation of a parent note, or for auto captions that aid deaf readers (Russell & Norvig, 2020). Tips may pull users deeper, and chatbots can answer questions in a tone. It also seems to bridge language gaps, making workflows smoother and allowing creators to reach more fans; however, quality could still wobble (Ryan & Stahl, 2021).

2.3.5.1.1 Content Democratization

GenAI seems to be changing the way media is created. Previously, making good videos or music required pricey gear, special training, and expensive studio software. Now, thanks to GenAI tools, a teenager with just a laptop can create something that looks like a professional product. Programs that edit videos, compose beats, or draw pictures are becoming easy to use; therefore, more people are joining the digital scene (Eloundou et al., 2023).

This shift may mean a more inclusive space, allowing groups that used to be left out to share their stories worldwide. However, some people worry that a flood of content could drown quality, and big companies might still steer trends behind the scenes. However, the overall power balance appears less centered on a few studios and is more spread out. It could shift taste.

2.3.5.1.2 Language Inclusivity

Language has traditionally been a significant barrier to international communication, often limiting access to mass media, learning, and political discussions. AI-based translation technologies, such as multilingual text generators and automatic real-time speech translation systems, are reducing these barriers to some extent (Stix, 2021).

By facilitating inter-linguistic interactions, GenAI encourages global deliberative inclusiveness and cultural exchange (Bahdanau, Cho, & Bengio, 2015). For example, a South American journalist can instantly translate news articles into other languages, thus reaching a global audience. Similarly, grassroots movement activists can overcome linguistic boundaries and thus strengthen transnational mutual action. These technologies are particularly promising for areas such as learning, international diplomacy, and joint crises spanning international borders (Morley 2020).

2.3.5.1.3 Crisis Communication

In emergencies, such as natural catastrophes, pandemics, or armed conflicts, the quick and correct dissemination of information is crucial. GenAI can assist tremendously through the preparation of real-time synopses of developing incidents, abstracting crucial data from massive data flows, and converting these outcomes into an assortment of languages (Zhang & Lu, 2021). For instance, AI-based systems can process social media content during hurricanes, discern affected zones, expedite relief delivery, and broadcast crucial safety messages. The utilization of GenAI considerably enhances disaster preparedness and response, where human communicators cannot cope with the volume or tempo of the information needed.

2.3.5.2 Challenges and Risks of GenAI in Media and Communication

GenAI may shake up the media, but it also poses risks. One concern is the spread of misinformation and deepfakes, which could twist public opinion and erode trust in content. Bias in data might keep stereotypes alive, shaping stories in harmful ways. At the same time, significant ownership questions arise because AI output blurs who wrote something. Relying too much on machines could dull human creativity and weaken the journalistic standards. Ethical issues—such as consent, privacy, and manipulation—further cloud the picture. Without safeguards, GenAI is likely to amplify disinformation, bias, and exploitation, challenging credibility and democratic discourse (Muller, 2020).

2.3.5.2.1 Misinformation and Deepfakes

The tools that enable mass-level content production also empower nefarious actors to create credible disinformation. Deepfakes—virtually indistinguishable from real visual, audio, or video impressions—pose a special problem because they blur the line between real and constructed media. Deepfakes are used to spread propaganda, blackmail influential figures, or distort political events, and they compromise the epistemic foundations of deliberative democracy (Chesney & Citron, 2019). For example, AI-created videos of politicians making false statements can destroy institutional credibility and influence election results. In addition, GenAI can create high volumes of fabricated news articles and overwhelm fact-checkers while amplifying disinformation campaigns. These threats necessitate verification mechanisms, regulations, and digital literacy awareness among the general populace.

2.3.5.2.2 Erosion of Journalism

Journalism has long been a cornerstone of democratic accountability, with investigative reporting revealing corruption, human rights abuses, and systematic injustice. Widening news production automation with GenAI might further lower journalism standards (Pasquale, 2015). While AI-assembled reporting might generate economic news updates or sports scores cheaply and efficiently, it cannot approximate the contextual judgment, critical questioning, and ethical responsibility that define investigative reporting (Diakopoulos, 2019).

Over-reliance on automated reporting might induce news institutions to economize with fewer human journalists and further weaken the watchdog role of the press. Such weakening could prove disastrous at the societal level because the press serves as an important check on power.

2.3.5.2.3 Information Overload

One problem seems to be the surge of fake stuff that keeps spilling over the media. GenAI can crank out tons of blogs, short clips, and even Tik-Tok posts in seconds. That flood may provide extra information, but it also clouds the truth. People often cannot tell a reliable news page from a clickbait meme (Helbing et al., 2021). Therefore, the "attention economy" drags people into constant data and tests their sense of reality.

Slowly, fatigue builds, and some people might stop scrolling. Moreover, messengers can help false stories spread quickly. In short, the boost is not all good for today. It may also undermine trust.

2.3.5.3 Navigating the Opportunities and Risks

GenAI's spread through news feeds and social sites brings both hope and worry. It appears to let more folks join worldwide chats, because it can turn jokes, memes, even tiny slogans into many tongues. Yet, false whispers may travel too perhaps. This could make crises feel less sudden and help societies bounce back. However, the same tools also spread false stories quickly, and they can wear down the habit of checking facts. Overload of AI-filled posts may make audiences numb, so the whole information scene feels shaky (Shneiderman, 2020). To curb these problems, a few steps may be useful.

- Laws that force clear tags on any AI-made clip or article, so readers see a warning sign.
- Technical tricks, such as digital watermarks or crypto stamps, may help spot fake pieces, although they are not perfect.
- School programs that teach kids to hunt clues, question sources, and not trust everything instantly could encourage skepticism.
- The rules for development studios and newsrooms should preserve democratic values, not destroy them.

GenAI can lift creative blocks, which may help struggling artists. It also allows unheard groups to share ideas, which is a hopeful sign. Quick alerts during crises This technology can handle this quickly. However, the risk of false news rises, and journalism standards can slip. Synthetic picture crowd feeds confuse viewers.

The future depends on community choices. Smart rules, safeguards, and basic lessons could guide AI toward open, fair talk while reducing bad effects. We must also monitor the power in corporate hands. In conclusion, the goal is not to stop AI technology but to guide it toward people-first outcomes.

2.3.6 Creative Industries

The worlds of art, music, books, movies, and design seem to be changing rapidly because of GenAI. It may provide creators with new ways to express their ideas, and tools appear easier for more people to use. Therefore, there could be more opportunities to earn money (Smith & Anderson, 2019). However, it also raises concerns about copyright, jobs, and keeping cultural variety alive globally.

2.3.6.1 Advantages of GenAI in Creative Industries

2.3.6.1.1 New Forms of Art

The biggest buzz about GenAI is how it sparks fresh art. Artists work side by side with AI to paint, write songs, and draft stories that bend the old ideas of creativity

(Elgammal et al., 2017). These teams often create pieces that mix human aim with machine output, showing looks that were not possible before. For example, AI-grown paintings have appeared in galleries.

In addition, code-crafted music now joins experimental shows. Some critics think this blend could dilute true expression, but many feel it expands the field and inspires students overall.

2.3.6.1.2 Lowering Barriers to Creativity

GenAI software places art production in the hands of people with software that enables non-professionals to create high-end work. From producing illustrations through film editing to composing music, one can engage in mass creation without learning technical software or professional artistic skills (Vinuesa et al., 2020).

Access to cultural production increases, and a much broader spectrum of stakeholders beyond experts can become involved in the process of creation. Consumers of art are no longer relegated to that position and can become involved in production.

2.3.6.1.3 Market Expansion

The commoditization of generative art tools has established new markets and income streams for artists. Online marketplaces, such as those housing non-fungible tokens (NFTs), allow artists to profit from AI-augmented works and global audiences with no middlemen. Similarly, AI-augmented design software helps small companies and solo practitioners become major competitors in creative markets once controlled by mega-studios. By cutting middleman expenses while extending reach, GenAI encourages entrepreneurial activities and acts as a catalyst for the creative economy (Arrieta, 2020).

2.3.6.2 Challenges and Risks of GenAI in Creative Industries

2.3.6.2.1 Intellectual Property Disputes

The use of GenAI in creative production raises the number of unanswered questions regarding authorship, ownership, and copyright. If an artwork is produced with extensive machine input, it becomes uncertain whether the credit should remain with the human operator or with the AI creator (or both). Such confusion defies the legitimate protection of artists and could destroy innovation incentives. Samuelson (2023) suggests that intellectual property systems are compelled to evolve and handle originality and attribution issues as AI-created works proliferate.

2.3.6.2.2 Job Displacement

Automation in the arts could bring economic concerns. Illustrators, designers, writers, and musicians may see demand shrink as AI outputs are used more for cheapness and speed. Therefore, firms might replace human work with machines. However, some jobs may appear in curation, editing, or AI training. These roles are unlikely to fully compensate for the loss of creative tasks. This shift raises questions about the value of human art. When a machine can pump out content, does it cheapen its value? This is a significant dilemma (O'Neil, 2016).

2.3.6.2.3 Cultural Homogenization

One problem may be that generative models often copy patterns that mainly originate from the majority culture. If most training data are taken from Western news or popular media, the AI's output could look less varied and push aside minority traditions.

Over time, this could lead to a sort of cultural homogenization, where global art looks the same. To keep creative work diverse, training sets should include many different cultural practices and identities.

New AI tools are changing the creative field, making it easier for more people to create art and opening new markets. However, questions remain regarding the law, job loss, and culture. Some argue that the risks might be overstated, while others see genuine danger. In a balanced future, policies need to protect intellectual property, help artists, and ensure that datasets remain inclusive. Otherwise, progress could end up trimming down our cultural richness today for the next generation worldwide.

By adopting such practices, societies can use GenAI as an enabling collaboration aid rather than a danger that weakens human creativity. Societies can leverage GenAI as a shaper of cultural domains through collaboration with AI.

2.3.7 Scientific Research

GenAI is now being used in laboratories, and it may change the way how scientists create and check ideas. From sparking new hypotheses to running huge simulations, the tool seems to speed up discovery and teamwork. However, there is a risk that it could hurt research honesty, and that risk is balanced by ethical rules and regulations (NIST, 2022). Therefore, the field feels both hopeful and uneasy in future research.

2.3.7.1 Advantages of GenAI in Scientific Research

2.3.7.1.1 Hypothesis Generation

One idea about GenAI in science is to find hidden patterns in huge datasets. This may involve observing trends in genes, chemicals, or weather that people may overlook. The system can suggest new research paths, although researchers might not spot them immediately. Thus, experiments provide new clues, while classic lab work remains relevant. This mix of data ideas and old methods might push discovery forward, even if the exact value is still uncertain.

2.3.7.1.2 Simulation and Modeling

GenAI might improve the efficacy of simulations in many fields, such as climate studies, material design, and physics. For example, it can predict how molecules arrange themselves or copy complex system behavior while using far less computer power than previous methods (Noé et al., 2019). This allows scientists to conduct experiments that would be too expensive, time-consuming, or impossible to conduct in a lab. However, some worry that shortcuts could hide errors that real tests would catch. Therefore, having more conditions to explore seems to open new doors for experimental work.

2.3.7.1.3 Collaboration Acceleration

Perhaps GenAI merely assists in finding information, but it also seems to encourage collaboration. Tools that write together can draft reviews, produce reports, and even fix the layout, so scientists have less paperwork.

This might free them to think more deeply about experiments. Moreover, these tools could bring far-off labs closer, costing less for new scholars and schools with tight budgets. Therefore, collaboration may significantly improve the overall research quality.

2.3.7.1.4 Case Study: AI in Drug Discovery

One clear case shows how GenAI may shape medicine, especially drug discovery. Insilico Medicine used generative models for pulmonary fibrosis. Their AI designed a molecule named INS018_055 and it reached pre-clinical testing in under-18 months. Normally this stage drags on for years, so the speed appears surprising (Zhavoronkov et al., 2019). Researchers think AI may speed up development trimming cost.

Some scientists argue it might open paths to illnesses that lack good candidates. Yet critics whisper that the model's predictions still need human checks. Results still suggest that AI could cut and money, yet uncertainties remain. Future trials may reveal whether the approach truly outperforms traditional ways significantly. Overall, the example hints at big potential in reality.

2.3.7.2 Challenges and Risks of GenAI in Scientific Research

2.3.7.2.1 Fabrication of Research Data

Modern tools that push science forward can also be used to craft fake datasets. Bad actors might use GenAI to spin counterfeit but believable results, which could shake confidence in research results. This suggests that peer review and data checks are becoming more urgent to keep journals trustworthy.

2.3.7.2.2 Publication Bias

One worry is the flood of AI-written papers in journals. If this continues, standards could slip, and real findings could be hidden. The bias in publishing may end up filling research with useless noise that bothers scholars and reviewers. Some say stricter rules might help, yet making them work is very difficult. Perhaps universities should teach students to spot fake content, though not all agree that this solves the issue today (The Royal Society, 2019).

2.3.7.2.3 Ethical Oversight

The rapid rise of GenAI tools in risky fields such as gene work, brain studies, and biotechnology raises significant moral questions. Regulation may fall behind what technology can do, so harmful uses may appear early. Without clear rules, GenAI research could cross ethical lines and hurt groups or people. The system could still shift science by providing fresh ideas, faster simulations, and better teamwork. However, there is a risk of fake data, slipping quality in papers, and weak oversight, which probably means careful use is not optional.

Setting open standards, proof steps, and rules may help the community keep GenAI a boost, not a sink, for trustworthy knowledge. Some scientists fear that excitement could hide hidden bias, which might skew results later. Researchers are concerned about the impact on education.

2.3.8 Comparative Synthesis

Industries may gain efficiency, personalized services, easier access, and new ideas from GenAI. However, the risks could include bias, false information, privacy loss, job cuts, and weak accountability.

The net effect depends on governance: benefits tend to prevail over risks in regulated environments, and harms will multiply in unregulated areas. A comparative overview is presented in Table 2.1.

The advantages and disadvantages of adopting GenAI depend highly on the context. Education apps collect a lot of data, which may mean learning but also raises privacy concerns.

Health portals tailor advice, but they appear to tip the equity balance.

In music streaming, anyone can share songs, but owners argue over rights, and workers feel insecure.

Law groups and cyber teams face a puzzle: they try to protect people while shaking up institutions. Companies should use a plan that adds fairness to the design, people watching the system, and sharing. Ethical guessing, rules, and teamwork across fields could prevent AI from hurting more than helping society.

TABLE 2.1
Pros and Cons of GenAI Across Industries

Industry	Pros	Cons
Healthcare	Diagnostics, drug discovery, education	Privacy risks, bias, accountability
Finance	Fraud detection, personalization, forecasting	Credit bias, manipulation, cybersecurity
Education	Personalized learning, accessibility, automation	Plagiarism, equity gap, depersonalization
Governance and Law	Policy simulation, legal assistance, citizen services	Transparency issues, bias, surveillance
Media	Content democratization, translation, crisis response	Deepfakes, erosion of journalism, overload
Creative Arts	New art forms, accessibility, markets	IP disputes, job loss, homogenization
Science	Hypothesis generation, modeling, collaboration	Fabrication, low-quality output, weak oversight

2.4 ETHICAL IMPLICATIONS ACROSS INDUSTRIES

The moral framework for GenAI may not be consistent; it shifts with each use case, industry demand, and what society thinks is right. Even if fairness, responsibility, clarity, and privacy appear to be universal rules, their real-world use tends to differ significantly across fields (Floridi & Cowls, 2019; Jobin et al., 2019). Therefore, this section examines how GenAI ethics play out in particular sectors, pointing out both possible benefits and risks that might follow for businesses and everyday users alike.

2.4.1 Healthcare and Biomedicine

GenAI might reshape healthcare, from how doctors spot diseases to how new drugs are made. Some scientists believe that fake medical pictures can help when real scans are few (e.g., a small hospital in Ohio). However, concerns still linger.

- *Patient Privacy*: Using real health records to teach computers raises questions about consent and actual hidden identities.
- *Synthetic Data Risk*: Even scrubbed data could still reveal a person if re-identification is successful.
- *Bias and Fairness*: Models built mostly on Western data may work less well for underserved groups, possibly widening inequities (Rajkomar et al., 2018).

Therefore, while fake datasets promise broader inclusion and faster research, we likely need strong significant safeguards for data governance in practice and algorithmic fairness overall.

2.4.2 Finance and Banking

GenAI models are increasingly embedded in fraud detection, risk modeling, and one-to-one financial counseling. The ability of models to replicate market settings or create artificial transaction data facilitates robust, stress testing. However, ethical questions arise.

- *Bias in Lending*: Generative models may exhibit structural bias in loan disapprovals (Hurley & Adebayo, 2016).
- *Fraudulent Deepfakes*: Synthetic voice and video can be weaponized for financial scams and identity theft, undermining institutional trust (West, 2019).
- *Transparency*: Black-box GenAI decision-making raises accountability concerns when customers are denied credit.

Regulators like The US Federal Reserve and the European Central Bank (ECB) have already emphasized explainability and accountability in AI-powered finance (European Central Bank, 2021).

2.4.3 Education and Academia

In learning, GenAI has the potential for one-to-one tutoring, grading by computers, and content creation. However, ethical issues are multifaceted.

- *Academic Integrity*: Students may misuse tools such as ChatGPT to produce essays or assignments, raising plagiarism concerns (Cotton et al., 2023).
- *Equitable Access*: Wealthier universities have been able to embed advanced GenAI systems, thus widening the digital divide.
- *Critical Thinking*: Many students believe that too many AI tutors may block creativity and analysis. So it is not just "Will AI be used?" but "how we mix tools while keeping human choices intact maybe" for students in education.

2.4.4 Creative Industries and Media

GenAI seems to be reshaping art, journalism, film, and music, allowing creators to make things instantly (McCormack et al., 2019). Some ethical concerns were noted:

- *Authorship and Copyright*: It is unclear who actually owns AI-created pieces—the programmers, the data providers, or the person who clicks "generate" (Samuelson, 2023).
- *Cultural Homogenization*: Relying heavily on popular datasets might push cultures toward the same look, hurting true variety.
- *Deepfakes and Truth Decay*: Manipulation of media by GenAI erodes public trust and democratic deliberation (Chesney & Citron, 2019). While GenAI has the ability to democratize creativity through the enablement of non-artists, it also poses challenges to traditional creative labor markets and laws.

2.4.5 Governance, Law, and Security

Governments and courts are looking at using GenAI to draft enforceable papers, predict crimes before they happen, and run public services (Završnik, 2019). Those ideas feel risky.

- *Civil Liberties Could Be Squeezed*: Predictive policing may lock certain races into extra watch.
- *Disinformation Spreads Rapidly*: When state or rebel groups push AI-made propaganda (Brundage et al., 2018).
- *Legal Accountability Becomes Fuzzy*: If an AI-written brief errs, who really holds the blame?

Therefore, human-rights rules and democratic checks should shape any rollout, stressing oversight and clear responsibility. In conclusion, a careful study seems unavoidable before adoption.

2.4.6 Cross-Sectorial Synthesis

Talking about AI ethics gets messy; bias and fairness could keep unfairness alive; we hardly see how AI makes choices, so transparency and accountability feel fuzzy; data privacy and consent matter a lot in hospitals or banks where records sit; and misinformation and trust crash when fake videos or false posts are spread.

Jobs and creativity are shaken as technology pushes people out or reshapes work; each field perceives these risks in its own way; some believe that a single rule works, but this probably ignores real differences—for example, fake news hit politics hard, while clinics worry about leaking patient information; knowing this may push lawmakers toward rules that fit each sector, not a blanket plan overall (IEEE, 2019).

2.5 CONCLUSION

GenAI feels like the biggest tech buzz right now. It can shoot out text, pictures, software, and even research models in moments. That speed may bring hope, yet also fear, across every field. Hope? Maybe new ideas bloom fast, but uncertainty lingers too for students and researchers everywhere today. GenAI enables faster discoveries, empowers more people to create, and makes workplaces operate more efficiently.

The worry part? Too-much info spreading, experts losing their voice, old divides getting louder. Whether GenAI ends up as a bright force or a dark one will depend on the rules and morals that we set up right now.

The ethical side of GenAI already shows cracks that make people nervous. In places where mistakes can cost lives—like police data tools, bank risk scores, or doctor diagnoses—folks worry about bias, fairness and the fact we cannot always see what the AI is doing. Surveys say these flaws shake trust in algorithms (Mehrabi et al., 2021). Also, the way GenAI spreads faster than privacy rules, rules for ownership, or ways to blame anyone, leaves lawmakers running after it (Crawford, 2021). If we ignore that, the tech could spread fake news, take jobs, make rich people richer, and turn a helpful tool into a weapon.

Still, the upside of GenAI looks huge. In a lab, a model can guess how a molecule will work, shaving years off drug research. Climate scientists use GenAI to run detailed atmosphere simulations, giving clearer pictures of how the planet might shift. Accessibility apps now have text-to-speech or image-description tools that help a student with dyslexia read a textbook, or let a blind person hear the layout of a room. Researchers have shown these uses can tackle some of humanity's toughest problems (Esteva et al., 2021). To keep new technology useful, moral watching is required; data should be fair, although the results may be messy; explanations must be readable by people, not just scientists; rules must combine law, ethics, and public input; and the future path is not fixed; it leans on the choices we make now.

Good governance means that many voices, makers, lawmakers, scholars, and users share the job, which does not halt invention but should steer it toward the public societal good; therefore, power must be aligned with democracy and fairness.

By building safety steps right into how we design and roll out GenAI, we can turn a risky guess into a reliable tool for everyone.

The moral charge for GenAI has two parts: it must stay a spring of creativity and efficiency while echoing the values we all care about. Without solid oversight, the tech will magnify the worst bits of us—prejudice, power gaps, and torn communities. Think of the ethical arena as a wild field we are mapping; every choice, from how an algorithm is built to which law passes, pushes the tech toward justice or toward clash. Therefore, stewardship is not an after-thought; it is the engine that decides where GenAI goes.

In short, the fate of GenAI sits in our collective hands. To keep it from hurting us, we need nonstop moral checking, a focus on human-centered values, and equal chance to use it. We are at a crossroads for humanity; how we watch, rule, and share this tech will say if it becomes a booster for progress or a source of trouble. By taking a stand for fairness, openness, and shared care now, societies can guide a future where GenAI lifts us all instead of pulling us apart.

EXERCISES

1. What are the essential advantages of GenAI for societal development and other industries?
2. How would you use GenAI to address intricate issues involving healthcare, education, and the creative industries?
3. What are the most important risks or unforeseen side effects related to GenAI?
4. How would it be feasible to achieve equilibrium between fostering creativity and imposing rules in development of GenAI?
5. What frameworks exist to assess the social impacts of GenAI, ethical issues, and considerations?
6. What are the substantive ethical considerations linked to the use of GenAI?
7. In what ways does GenAI defy accepted thinking about authorship, ownership, and intellectual property rights?
8. What are the consequences of generative model biases in AI, and how can they be mitigated through adequate measures?
9. How could GenAI jeopardize individuals' privacy and autonomy?
10. How might ethical principles for AI be incorporated into the building of creative AI?
11. Describe benefits and flaws of implementing AI across industries and sectors.
12. How much does the use of GenAI vary across industries such as healthcare, financial services, education, and media?
13. What are the top benefits that accrue to firms from adopting GenAI technologies?
14. What are the possible negative or ethical implications unique to industries with GenAI?
15. How can industries implement effective measures to ensure the responsible and ethical use of GenAI?

REFERENCES

Arrieta, A. B., Díaz-Rodríguez, N., Del Ser, J., Bennetot, A., Tabik, S., Barbado, A., . . . Herrera, F. (2020). Explainable Artificial Intelligence (XAI): Concepts, taxonomies, opportunities and challenges toward responsible AI. *Information Fusion*, *58*, 82–115. https://doi.org/10.1016/j.inffus.2019.12.012

Bahdanau, D., Cho, K., & Bengio, Y. (2015). Neural machine translation by jointly learning to align and translate. *International Conference on Learning Representations (ICLR).* https://arxiv.org/abs/1409.0473

Barocas, S., Hardt, M., & Narayanan, A. (2019). *Fairness and machine learning.* fairmlbook.org. http://fairmlbook.org

Bender, E. M., Gebru, T., McMillan-Major, A., & Mitchell, S. (2021). On the dangers of stochastic parrots: Can language models be too big? *Proceedings of the 2021 ACM Conference on Fairness, Accountability, and Transparency (FAccT '21)*, 610–623. https://doi.org/10.1145/3442188.3445922

Boddington, P. (2017). *Towards a code of ethics for artificial intelligence.* Springer. https://doi.org/10.1007/978-3-319-60648-4

Brundage, M., Avin, S., Clark, J., Toner, H., Eckersley, P., Garfinkel, B., Dafoe, A., et al. (2018). The malicious use of artificial intelligence: Forecasting, prevention, and mitigation. *arXiv.* https://arxiv.org/abs/1802.07228

Brynjolfsson, E., & McAfee, A. (2017). *Machine, platform, crowd: Harnessing our digital future.* W. W. Norton & Company.

Bryson, J. J. (2018). Patiency is not a virtue: The design of intelligent systems and systems of ethics. *Ethics and Information Technology*, *20*(1), 15–26. https://doi.org/10.1007/s10676-018-9448-6

Burrell, J. (2016). How the machine 'thinks': Understanding opacity in machine learning algorithms. *Big Data & Society*, *3*(1), 1–12. https://doi.org/10.1177/2053951715622512

Cath, C. (2018). Governing artificial intelligence: Ethical, legal and technical opportunities and challenges. *Philosophical Transactions of the Royal Society A*, *376*(2133), 20180080. https://doi.org/10.1098/rsta.2018.0080

Carlini, N., Tramer, F., Wallace, E., Jagielski, M., Herbert-Voss, A., Lee, K., Roberts, S., Brown, T., Song, D., Erlingsson, Ú., & Ristenpart, T. (2021). Extracting training data from large language models. *Proceedings of the 2021 IEEE Symposium on Security and Privacy (SP)* (pp. 1–18). IEEE. https://doi.org/10.1109/SP40001.2021.00041

Chenthamarakshan, V., Das, P., Hoffman, S. C., Strobelt, H., Padhi, I., Lim, K. W., Hoover, B., Manica, M., Mojsilović, A., & Ramamurthy, K. (2020). CogMol: Target-specific and selective drug design for COVID-19 using deep generative models. *arXiv preprint arXiv:2004.01215.*

Chesney, R., & Citron, D. K. (2019). Deep fakes: A looming challenge for privacy, democracy, and national security. *California Law Review*, *107*(6), 1753–1819. https://doi.org/10.2139/ssrn.3213954

Cho, K., Ahn, J., & Kim, S. (2023). Bias in text-to-image generation: Gender associations in AI visual outputs. *Proceedings of the 2023 ACM Conference on Fairness, Accountability, and Transparency (FAccT '23)* (pp. 1–13). ACM Digital Library. https://doi.org/10.1145/3593013.3594078

Citron, D. K. (2020). *The fight for privacy: Protecting dignity, identity, and love in the digital age.* W. W. Norton & Company.

Cotton, D. R. E., Nash, R. A., & Davies, C. (2023). The challenges of generative AI in higher education: Risks for academic integrity and learning. *Assessment & Evaluation in Higher Education*, *48*(4), 523–539. https://doi.org/10.1080/02602938.2023.XXXXXX

Crawford, K. (2021). *Atlas of AI: Power, politics, and the planetary costs of artificial intelligence.* Yale University Press. https://doi.org/10.12987/9780300252392

Creemers, R. (2022). *China's approach to AI governance: National strategies and regulatory frameworks.* China Governance Project, University of Oxford.

Diakopoulos, N. (2019). *Automating the news: How algorithms are rewriting the media.* Harvard University Press. https://doi.org/10.4159/9780674242988

Dignum, V. (2019). *Responsible artificial intelligence: Developing and using AI in a responsible way.* Springer. https://doi.org/10.1007/978-3-030-30371-6

Elgammal, A., Liu, B., Elhoseiny, M., & Mazzone, M. (2017). CAN: Creative adversarial networks, generating "art" by learning about styles and deviating from style norms. *Proceedings of the 8th International Conference on Computational Creativity (ICCC)*, 96–103.

Elkins, J. (2023). AI and the art market: Challenges for artists in the age of generative tools. *Art Journal*, *82*(2), 45–62. https://doi.org/10.1080/00043249.2023.XXXXXX

Eloundou, T., Manning, S., Mishkin, P., & Rock, D. (2023). GPTs are GPTs: An early look at the labor market impact potential of large language models. *OpenAI Research.* https://arxiv.org/abs/2304.09103

Esteva, A., Robicquet, A., Ramsundar, B., Kuleshov, V., DePristo, M., Chou, K., Cui, C., Corrado, G., Thrun, S., & Dean, J. (2021). A guide to deep learning in healthcare. *Nature Medicine*, *25*(1), 24–29.

Eubanks, V. (2018). *Automating inequality: How high-tech tools profile, police, and punish the poor.* St. Martin's Press.

European Central Bank. (2021). *Artificial intelligence and machine learning in financial services: Supervisory expectations and accountability.* European Central Bank.

European Commission. (2021). *Proposal for a regulation laying down harmonised rules on artificial intelligence (Artificial Intelligence Act) and amending certain Union legislative acts.* Publications Office of the European Union.

Feldstein, S. (2019). The global expansion of AI surveillance. *Carnegie Endowment for International Peace.*

Floridi, L. (2019). Establishing the rules for building trustworthy AI. *Nature Machine Intelligence*, *1*(6), 261–262. https://doi.org/10.1038/s42256-019-0055-y

Floridi, L., & Cowls, J. (2019). A unified framework of five principles for AI in society. *Harvard Data Science Review*, *1*(1). https://doi.org/10.1162/99608f92.8cd550d1

Gervais, D. J. (2020). The machine as author. *Iowa Law Review*, *105*(5), 2053–2106.

Gillespie, T. (2018). *Custodians of the internet: Platforms, content moderation, and the hidden decisions that shape social media.* Yale University Press.

Goodfellow, I. J., Pouget-Abadie, J., Mirza, M., Xu, B., Warde-Farley, D., Ozair, S., Courville, A., & Bengio, Y. (2014). Generative adversarial networks. In *Advances in neural information processing systems* (Vol. 27). Curran Associates, Inc.

Hagendorff, T. (2020). The ethics of AI ethics: An evaluation of guidelines. *Minds and Machines*, *30*(1), 99–120. https://doi.org/10.1007/s11023-020-09517-8

Hao, K. (2020, February 4). We read the paper that forced Timnit Gebru out of Google: Here's what it says. *MIT Technology Review.* https://www.technologyreview.com/2020/12/04/1013294/ai-ethics-paper-forced-timnit-gebru-google/

Heaven, W. D. (2020, April 21). AI could help with the next pandemic—but not with this one. *MIT Technology Review.* https://www.technologyreview.com/2020/04/21/1000657/ai-could-help-with-the-next-pandemic-but-not-with-this-one/

Helbing, D. (2019). *Towards digital enlightenment: Essays on the dark and light sides of the digital revolution.* Springer. https://doi.org/10.1007/978-3-319-90869-4

Helbing, D., Frey, B. S., Gigerenzer, G., Hafen, E., Hagner, M., Hofstetter, Y., . . . Zwitter, A. (2021). Will democracy survive big data and artificial intelligence? *Scientific American*, *27*(1), 1–9.

Hern, A. (2019, January 25). Chatbots are a danger to democracy. *The Guardian*. https://www.theguardian.com/technology/2019/jan/25/chatbots-are-a-danger-to-democracy

Huang, G. (2020, September 14). The state of AI in 2020. *MIT Sloan Management Review*. https://sloanreview.mit.edu/article/the-state-of-ai-in-2020/

Hurley, M., & Adebayo, J. (2016). *Algorithmic bias in financial services: Challenges and opportunities. Springer briefs in ethics and information technology*. Springer. https://doi.org/10.1007/978-3-319-41479-0

IBM. (2020). Principles for trust and transparency. *IBM Research*. https://www.ibm.com/blogs/research/2020/01/principles-for-trust-and-transparency

IEEE. (2019). *Ethically aligned design: A vision for prioritizing human well-being with autonomous and intelligent systems* (1st ed.). IEEE. https://ethicsinaction.ieee.org

Jobin, A., Ienca, M., & Vayena, E. (2019). The global landscape of AI ethics guidelines. *Nature Machine Intelligence*, *1*(9), 389–399. https://doi.org/10.1038/s42256-019-0088-2

Jumper, J., Evans, R., Pritzel, A., Green, T., Figurnov, M., Ronneberger, O., Tunyasuvunakool, K., Bates, R., Žídek, A., Potapenko, A., Bridgland, A., Meyer, C., Kohl, S. A. A., Ballard, A. J., Cowie, A., Romera-Paredes, B., Nikolov, S., Jain, R., Adler, J., . . . DeepMind. (2021). Highly accurate protein structure prediction with AlphaFold. *Nature*, *596*(7873), 583–589. https://doi.org/10.1038/s41586-021-03819-2

Kaissis, G. A., Makowski, M. R., Rückert, D., & Braren, R. F. (2020). Secure, privacy-preserving and federated machine learning in medical imaging. *Nature Machine Intelligence*, *2*(6), 305–311. https://doi.org/10.1038/s42256-020-0186-1

Kenderdine, S. (2020). Deep fakes: A critical perspective on AI and cultural heritage. *ACM International Conference on Interactive Experiences for TV and Online Video (TVX)*. Association for Computing Machinery.

Kerr, I. (2019). Prediction, pre-emption, presumption: How big data threatens big picture privacy. *Stanford Law Review Online*, *66*, 65–72.

King, R. D., Rowland, J., Oliver, S. G., Young, M., Aubrey, W., Byrne, E., Liakata, M., Markham, M., Pir, P., & Soldatova, L. N. (2020). The automation of science. *Science*, *324*(5923), 85–89. https://doi.org/10.1126/science.1165620

Krafft, P. M., Young, M., Katell, M., Huang, K., & Bugingo, G. (2020). Defining AI in policy versus practice. *Proceedings of the AAAI/ACM Conference on AI, Ethics, and Society*, 72–78. https://doi.org/10.1145/3375627.3375808

Leslie, D. (2019). Understanding artificial intelligence ethics and safety: A guide for the responsible design and implementation of AI systems in the public sector. *The Alan Turing Institute*. https://doi.org/10.5281/zenodo.3240529

Lin, P., Abney, K., & Bekey, G. A. (Eds.). (2017). *Robot ethics 2.0: From autonomous cars to artificial intelligence*. Oxford University Press. https://doi.org/10.1093/oso/9780190652951.001.0001

Lipton, Z. C. (2018). The mythos of model interpretability. *Communications of the ACM*, *61*(10), 36–43. https://doi.org/10.1145/3233231

Lu, Y. (2019). Artificial intelligence: A survey on evolution, models, applications and future trends. *Journal of Management Analytics*, *6*(1), 1–29. https://doi.org/10.1080/23270012.2019.1570365

Luckin, R. (2018). *Machine learning and human intelligence: The future of education for the 21st century*. UCL IOE Press.

Luckin, R., Holmes, W., Griffiths, M., & Forcier, L. B. (2016). *Intelligence unleashed: An argument for AI in education*. Pearson Education.

Marconi, F., & Siegman, A. (2017). *The future of augmented journalism: A guide for newsrooms in the age of smart machines*. Tow Center for Digital Journalism, Columbia University.

Marcus, G., & Davis, E. (2020). *Rebooting AI: Building artificial intelligence we can trust*. Princeton University Press.

McCormack, J., Gifford, T., Hutchings, P., & Zhu, J. (2019). Autonomous creativity: AI as a co-creator in the arts. *Leonardo*, *52*(3), 262–269. https://doi.org/10.1162/leon_a_01685

Mehrabi, N., Morstatter, F., Saxena, N., Lerman, K., & Galstyan, A. (2021). A survey on bias and fairness in machine learning. *ACM Computing Surveys*, *54*(6), 1–35.

Mittelstadt, B. D. (2019). Principles alone cannot guarantee ethical AI. *Nature Machine Intelligence*, *1*(11), 501–507. https://doi.org/10.1038/s42256-019-0114-4

Mittelstadt, B. D., Allo, P., Taddeo, M., Wachter, S., & Floridi, L. (2016). The ethics of algorithms: Mapping the debate. *Big Data & Society*, *3*(2), 1–21. https://doi.org/10.1177/2053951716679679

Moor, J. H. (2006). The nature, importance, and difficulty of machine ethics. *IEEE Intelligent Systems*, *21*(4), 18–21. https://doi.org/10.1109/MIS.2006.80

Morley, J., Floridi, L., Kinsey, L., & Elhalal, A. (2020). From what to how: An initial review of publicly available AI ethics tools, methods and research to translate principles into practices. *Science and Engineering Ethics*, *26*(4), 2141–2168. https://doi.org/10.1007/s11948-019-00165-5

Muller, V. C. (Ed.). (2020). *Philosophy and theory of artificial intelligence 2017*. Springer. https://doi.org/10.1007/978-3-319-96448-5

National Institute of Standards and Technology (NIST). (2022). *Artificial intelligence risk management framework (AI RMF 1.0)*. U.S. Department of Commerce. https://doi.org/10.6028/NIST.AI.100-1

Noble, S. U. (2018). *Algorithms of oppression: How search engines reinforce racism*. New York University Press.

Noé, F., Olsson, S., Köhler, J., & Wu, H. (2019). Boltzmann generators: Sampling equilibrium states of many-body systems with deep learning. *Science*, *365*(6457), eaaw1147. https://doi.org/10.1126/science.aaw1147

Obermeyer, Z., Powers, B., Vogeli, C., & Mullainathan, S. (2019). Dissecting racial bias in an algorithm used to manage the health of populations. *Science*, *366*(6464), 447–453. https://doi.org/10.1126/science.aax2342

O'Neil, C. (2016). *Weapons of math destruction: How big data increases inequality and threatens democracy*. Crown Publishing Group.

Pasquale, F. (2015). *The black box society: The secret algorithms that control money and information*. Harvard University Press. https://doi.org/10.4159/harvard.9780674736061

Rahwan, I. (2018). Society-in-the-loop: Programming the algorithmic social contract. *Ethics and Information Technology*, *20*(1), 5–14. https://doi.org/10.1007/s10676-017-9430-8

Rajkomar, A., Hardt, M., Howell, M. D., Corrado, G., & Chin, M. H. (2018). Ensuring fairness in machine learning to advance health equity. *Annals of Internal Medicine*, *169*(12), 866–872. https://doi.org/10.7326/M18-1990

Remus, D., & Levy, F. (2016). Can robots be lawyers? Computers, lawyers, and the practice of law. *Georgetown Journal of Legal Ethics*, *30*(3), 501–558.

The Royal Society. (2019). *Protecting privacy in practice: The current use, development and limits of privacy enhancing technologies in data analysis*. https://royalsociety.org/-/media/policy/projects/privacy-enhancing-technologies/privacy-enhancing-technologies-report.pdf

Russell, S., Dewey, D., & Tegmark, M. (2015). Research priorities for robust and beneficial artificial intelligence. *AI Magazine*, *36*(4), 105–114. https://doi.org/10.1609/aimag.v36i4.2577

Russell, S., & Norvig, P. (2020). *Artificial intelligence: A modern approach* (4th ed.). Pearson.

Ryan, M., & Stahl, B. C. (2021). Artificial intelligence ethics guidelines for developers and users: Clarifying their content and normative implications. *Journal of Information, Communication and Ethics in Society*, *19*(1), 61–86. https://doi.org/10.1108/JICES-12-2019-0138

Samuelson, P. (2023). Generative AI meets copyright. *Journal of the Copyright Society of the USA*, *70*(2), 197–246. https://doi.org/10.2139/ssrn.4360955

Shneiderman, B. (2020). Human-centered artificial intelligence: Reliable, safe & trustworthy. *International Journal of Human–Computer Interaction*, *36*(6), 495–504. https://doi.org/10.1080/10447318.2020.1741118

Smith, J., & Anderson, M. (2019, December 10). AI, robotics, and the future of jobs. *Pew Research Center*. https://www.pewresearch.org/internet/2019/12/10/ai-robotics-and-the-future-of-jobs/

Stix, C. (2021). Actionable principles for artificial intelligence policy: Three pathways. *AI & Society*, *36*(2), 533–555. https://doi.org/10.1007/s00146-020-00989-2

Strubell, E., Ganesh, A., & McCallum, A. (2019). Energy and policy considerations for deep learning in NLP. *Proceedings of the 57th Annual Meeting of the Association for Computational Linguistics* (pp. 3645–3650). Association for Computational Linguistics. https://doi.org/10.18653/v1/P19-1355

Surden, H. (2019). Artificial intelligence and law: An overview. *Georgia State University Law Review*, *35*(4), 1305–1336.

Vaithilingam, P., Zhang, T., & Glassman, E. L. (2022). Expectation vs. experience: Evaluating the usability of code generation tools powered by large language models. *Proceedings of the 2022 CHI Conference on Human Factors in Computing Systems*. Association for Computing Machinery. https://doi.org/10.1145/3491102.3517506

Vaswani, A., Shazeer, N., Parmar, N., Uszkoreit, J., Jones, L., Gomez, A. N., Kaiser, Ł., & Polosukhin, I. (2017). Attention is all you need. In *Advances in neural information processing systems* (Vol. 30). Curran Associates, Inc.

Vinuesa, R., Azizpour, H., Leite, I., Balaam, M., Dignum, V., Domisch, S., Felländer, A., Langhans, S. D., Tegmark, M., & Fuso Nerini, F. (2020). The role of artificial intelligence in achieving the sustainable development goals. *Nature Communications*, *11*, Article 233. https://doi.org/10.1038/s41467-019-14108-y

West, S. M. (2019). *Data-driven deception: The uses and abuses of synthetic media*. Center for Media, Data, and Society, NYU. https://datasociety.net/library/data-driven-deception/

Xie, T., France-Lanord, A., Wang, Y., Shao-Horn, Y., & Grossman, J. C. (2021). Crystal diffusion variational autoencoder for periodic material generation. *International Conference on Learning Representations (ICLR)*.

Yi, X., Walia, E., & Babyn, P. (2019). Generative adversarial network in medical imaging: A review. *Medical Image Analysis*, *58*, 101552. https://doi.org/10.1016/j.media.2019.101552

Yoon, J., Jarrett, D., & van der Schaar, M. (2020). Anonymization vs. personalization: Generating synthetic data with deep generative models. *Advances in Neural Information Processing Systems*, *33*, 13588–13600.

Whittlestone, J., Nyrup, R., Alexandrova, A., & Cave, S. (2019). The role and limits of principles in AI ethics: Towards a focus on tensions. *Proceedings of the AAAI/ACM Conference on AI, Ethics, and Society*, 195–200. https://doi.org/10.1145/3306618.3314289

Winfield, A. F. T., & Jirotka, M. (2018). Ethical governance is essential to building trust in robotics and artificial intelligence systems. *Philosophical Transactions of the Royal Society A*, *376*(2133), 20180085. https://doi.org/10.1098/rsta.2018.0085

Wirtz, B. W., Weyerer, J. C., & Geyer, C. (2019). Artificial intelligence and the public sector: Applications and challenges. *International Journal of Public Administration*, *42*(7), 596–615. https://doi.org/10.1080/01900692.2018.1498103

Završnik, A. (2019). Artificial intelligence in criminal law and public administration: Opportunities and challenges. *European Journal of Crime, Criminal Law and Criminal Justice*, *27*(3), 213–234.

Zellers, R., Holtzman, A., Bisk, Y., Farhadi, A., & Choi, Yejin. (2019). Defending against neural fake news. In *Advances in neural information processing systems* (Vol. 32). Curran Associates, Inc.

Zeng, Y., Lu, E., & Huangfu, C. (2019). Linking artificial intelligence principles. *arXiv*. https://doi.org/10.48550/arXiv.1812.04814

Zhang, C., & Lu, J. (2021). AI for disaster management: A survey. *AI Open*, *2*(1), 1–13. https://doi.org/10.1016/j.aiopen.2021.01.001

Zhavoronkov, A., Ivanenkov, Y. A., Aliper, A., Veselov, M. S., Aladinskiy, V. A., Aladinskaya, A. V., . . . Aspuru-Guzik, A. (2019). Deep learning enables rapid identification of potent DDR1 kinase inhibitors. *Nature Biotechnology*, *37*(9), 1038–1040. https://doi.org/10.1038/s41587-019-0224-x

Zuboff, S. (2019). *The age of surveillance capitalism: The fight for a human future at the new frontier of power*. PublicAffairs.

3 Exploring AI Learning Paradigms

From Supervised to Generative Learning

3.1 AI LEARNING PARADIGMS

The evolution of AI can be seen through the significant advancements in areas of ML, NLP, image recognition, etc. The AI system has evolved gradually from simple rule-based systems to dynamic areas truly powered by data, computation, and adaptive learning. The concept of learning paradigms plays a central role in the evolution of different AI technologies. The learning paradigms are the mechanisms through which knowledge is acquired, interpreted, and distilled into experience by machines. These paradigms define how an AI system perceives its environment, processes information, and generalizes from data to make decisions or generate new insights.

The different learning paradigms provide a framework that describes the diverse methodologies implemented in ML. Figure 3.1 shows how different learning paradigms are employed in AI. This chapter describes various learning paradigms along with their methodologies and application areas. The different learning paradigms introduced in this chapter are supervised, unsupervised, semi-supervised, reinforcement, self-supervised, and generative learning.

This chapter covers the paradigmatic continuum of AI learning, from conceptual roots and methodologies to applications. It provides the detailed study of how each paradigm contributes to the growing intelligence of today's AI systems: from pattern recognition and outcome prediction to generating realistic content and reasoning across modalities. By reading this, readers should come up with an appreciation for how traditional and emerging learning models interlink and collectively shape the next generation of intelligent, generative systems.

3.2 SUPERVISED LEARNING

Supervised learning is one of the most common approaches of ML in AI. In this approach, the model learns with labeled data. As shown in Figure 3.2, the AI system is trained on input data with known labels. The aim of the model is to map the input to correct outputs with the help of labels such that it makes accurate predictions when new information is presented (Sarker, 2021), for example, training a model to recognize the input image as "Cat" or a "Dog."

DOI: 10.1201/9781003713951-3

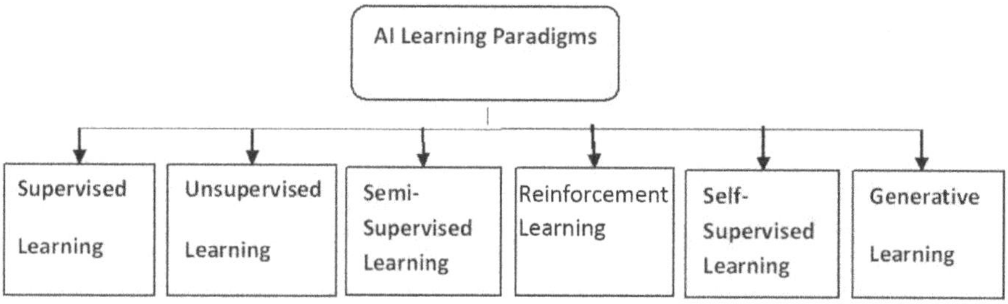

FIGURE 3.1 AI Learning Paradigms

FIGURE 3.2 Supervised Learning

It requires a lot of labeled data and careful training of the model to avoid overfitting. Despite its difficulties, supervised learning remains a pillar of modern AI development. The supervised learning is performed using various algorithms such as linear regression, logistic regression, decision trees, random forests, support vector machines (SVMs), and neural networks (including CNNs, RNNs).

It is commonly used in image recognition, sentiment analysis, fraud detection, and financial forecasting applications. The supervised learning is performed as classification and regression.

The classification trains the model to predict discrete labels or categories. Classification is extensively used in various industries where the decision-making is data-driven. It is extensively employed for different task in various sectors.

It utilizes machine learning models such as SVM, decision trees, and DL, namely CNNs, for medical diagnosis and disease detection using medical images; algorithms like logistic regression, random forest, and neural networks classify the transactions as either fraudulent or not, for detection of any fraud and for credit scoring; concept of NLP with Naïve Bayes, LSTMs, and Transformer models classify the customer reviews as positive or negative. Classification algorithms such as Naïve Bayes, decision trees, and DL are used in classifying spam emails from valid ones, for malware detection and spam filtering. They may also be utilized to classify objects of real-time captured images through CNNs and YOLO (You Only Look Once) model for object detection and traffic sign recognition.

In regression, the model predicts continuous numerical values, optimize operations, and improve decision-making. Regression is widely used in applications that require predicting numerical values. Linear regression, multiple regression, and gradient boosting are widely used in models to predict house prices; finance (stock price prediction and risk assessment) sector uses time series regression, LSTM neural networks, and ARIMA models in forecasting future stock prices; In healthcare, logistic regression is widely used for risk classification and polynomial regression for modeling disease progression. Multiple regression and XGBoost models are used to predict future sales. Linear regression and time series models can predict energy consumption patterns. The delivery time and traffic prediction models employ regression models, such as Lasso regression and random forest regression, in determining delivery time predictions.

Some of the key algorithms used for supervised ML are shown in Table 3.1 (Taye, 2023).

3.3 UNSUPERVISED LEARNING

Unsupervised learning is an ML field where models learn patterns, structures, or relationships from unlabeled data without predefined categories or outputs. As shown in Figure 3.3, the unsupervised learning finds hidden patterns from the given data without human intervention, for example, grouping customers by purchasing behavior without knowing their identity or preferences beforehand.

Table 3.2 shows different algorithms for unsupervised learning. Clustering, the easiest task in unsupervised learning, is used to group data points by similarity. Dimensionality reduction helps visualize high-dimensional data and improves efficiency in ML tasks. Generative models attempt to learn the data distribution to generate new samples that resemble the training data. Some of the key algorithms used for unsupervised ML are shown in Table 3.2 (Mahadevkar et al., 2022).

The unsupervised learning is applied in many applications. It is widely used for anomaly detection that identifies unusual or infrequent patterns in data. Clustering and density-based methods, such as k-means and DBSCAN, help identify outliers for financial fraud detection, network security, and industrial fault detection.

In business analytics and marketing, unsupervised learning helps in market segmentation by partitioning customers with similar buying habits, demographics, or tastes. k-means clustering and hierarchical clustering algorithms help businesses personalize marketing and provide personalized recommendations. Unsupervised learning helps in data compression by dimensionality reduction of high-dimensional data without losing valuable information.

PCA and autoencoders in DL are used extensively to achieve efficient data representation in image compression, speech processing, and storage optimization. Feature extraction is also performed using unsupervised learning. It is the process of transforming raw data into a more informative representation by identifying important patterns and structures. Methods like PCA, Independent Component Analysis (ICA), and DL-based autoencoders are used in image processing, NLP, and biomedical data analysis to enhance model performance.

TABLE 3.1
Supervised Machine Learning Algorithms

Algorithm	Type	Advantages	Limitations	Applications	References
Naïve Bayes	Classification	Fast, handles high-dimensional data	Assumes feature independence	Spam detection, text classification	Webb et al. (2010), Yang (2018)
Logistic Regression	Classification	Simple, interpretable	Not ideal for complex relationships	Medical diagnosis, fraud detection	LaValley (2008), Stoltzfus (2011)
Decision Tree	Both	Easy to interpret, handles all data types	Prone to overfitting	Loan approval, disease classification	Ying (2015), Quinlan (1996)
Random Forest	Both	Reduces overfitting, good accuracy	Less interpretable, slow	Stock prediction, medical imaging	Rigatti (2017), Breiman (2001)
SVM	Both	Works well in high dimensions	Hard to tune, slow with large data	Face recognition, bioinformatics	Suthaharan (2016), Pisner and Schnyer (2020)
k-NN	Both	Simple, no training needed	Slow, sensitive to noisy data	Handwriting recognition, recommendation	Guo et al. (2003), Zhang (2021)
Linear Regression	Regression	Simple, interpretable	Sensitive to outliers, assumes linearity	House price, salary prediction	James et al. (2023),
Ridge Regression	Regression	Reduces overfitting	Doesn't select important features	Stock analysis, multicollinear data	McDonald (2009)
Lasso Regression	Regression	Feature selection, good for sparse data	May miss correlated features	Feature selection, high-dimensional data	Ranstam and Cook (2018)
Gradient Boosting/ XGBoost	Both	High accuracy, handles missing data	Prone to overfitting, slow	Sales forecasting, risk assessment	Chen et al. (2015)

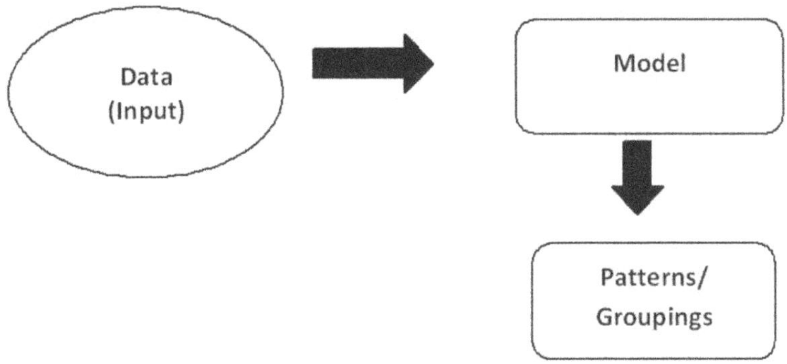

FIGURE 3.3 Unsupervised Learning Process Flow

3.4 SEMI-SUPERVISED LEARNING

Semi-supervised learning (SSL) is an ML approach that uses some labeled data with a lot of unlabeled data to enhance model performance. It fills the gap between supervised and unsupervised learning, so it is especially valuable when data labeling is costly or time-intensive.

The different algorithms for SSL are presented in Table 3.3. The basic idea of semi-supervised algorithm is shown in Figure 3.4. This process involves training a model first from few labeled datasets and iteratively using the same model to classify the un-labeled dataset to predict pseudo-labels.

The semi-supervised approach algorithms are self-training, co-training, and graph based. In self-training approach, the most confident predictions are iteratively added to the training set, improving the model. The co-training allows each model to produce pseudo-labels for the unlabeled samples, which are used by the other model for training, enhancing overall learning. The graph-based model approach labels propagate through the graph structure, so unlabeled nodes can inherit knowledge from labeled nodes efficiently.

SSL is employed for enhancing image segmentation and classification in medical imaging when acquiring labeled samples is expensive (e.g., tumor detection within MRI scans). In speech processing, SSL enables the training of models with a small amount of transcribed audio data, and it improves automatic speech recognition (ASR) systems. SSL helps to classify web pages, blog articles, and social media content using a small number of manually tagged data to categorize huge quantities of unlabeled information.

3.5 REINFORCEMENT LEARNING

Reinforcement learning (RL) is a paradigm of ML in which an agent learns to make decisions based on interactions with an environment (Mazyavkina et al., 2021). As shown in Figure 3.5, the agents perform actions after receiving feedback in terms of rewards or penalties and tries to maximize the cumulative reward over time, for

TABLE 3.2
Unsupervised Machine Learning Algorithms

Unsupervised Learning Algorithms		Method	Advantages, Limitations, and Research Focus	Applications
Clustering Algorithms	**K-Means Clustering**	It maps a data point to one of k-clusters by minimizing intra-cluster variance. It Implements Lloyd's algorithm: (i) Initialize cluster centroids, (ii) Assign points to the nearest centroid, (iii) Recalculate centroids, (iv) Repeat until convergence.	**Challenges:** Requires predefining k, sensitive to outliers and assumes spherical shape clusters. **Research Focus:** Can be improved with K-Means++ initialization or with deep learning (e.g., deep embedded clustering).	• **Anomaly detection** • **Market segmentation** • **Data compression** • **Feature extraction**
	DBSCAN (Density-Based Spatial Clustering of Applications with Noise)	Clusters points by density, identifying dense regions with sparse regions between them. Unlike K-Means, it does not need predefining k and can identify arbitrary-shaped clusters.	**Challenges:** Sensitive to hyperparameters ε(radius) and minPts (minimum points per cluster). **Research Focus:** Improvements include adaptive DBSCAN and hybrid DBSCAN-deep learning models.	
	Hierarchical Clustering	It constructs a dendrogram showing how data points are merged into clusters hierarchically. It is categorized into agglomerative approach (bottom-up merging) and divisive approach (top-down splitting).	**Advantages**: No need to predefine k, interpretable. **Limitations:** Computationally expensive for large datasets.	
Dimensionality Reduction Algorithms	**Principal Component Analysis (PCA)**	Projects data onto directions of maximum variance using eigenvectors of the covariance matrix. Mathematical basis: Solves the eigenvalue decomposition of the covariance matrix.	**Limitations:** Assumes linear relationships, sensitive to scaling.	
	Autoencoders (AEs)	Neural network-based approach to learn compressed representations. It consists of an encoder (compresses input to latent space) and a decoder (reconstructs input).	**Variants:** Denoising AEs, VAEs, Sparse AEs. **Research Focus:** Used in self-supervised learning, anomaly detection, and generative models.	

Generative Models	**Generative Adversarial Networks (GANs)**	Includes a generator (generates synthetic samples) and a discriminator (discriminates between synthetic and real). Training with adversarial loss: The generator improves by deceiving the discriminator.	**Challenges:** Mode collapse, instability, difficulty in training. **Research Focus:** Improved variants such as WGAN (Wasserstein GAN), StyleGAN, CycleGAN.
	Variational Autoencoders (VAEs)	Introduces probabilistic latent variables to capture data distribution. With a KL-divergence loss term to make the latent space follow a normal distribution.	**Comparison with GANs:** VAEs generate diversity but produce slightly blurry images. **Research Focus:** Hybrid models (e.g., VAE-GANs).

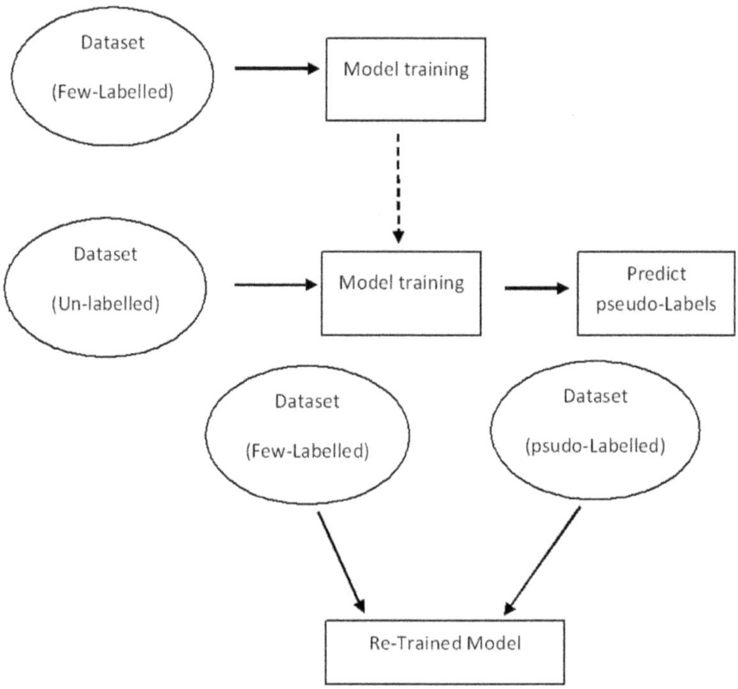

FIGURE 3.4 Semi-Supervised Learning Process Flow

TABLE 3.3
Semi-Supervised Machine Learning Algorithms

Method	Key Idea	Advantages	Limitations	Applications	References
Self-training	Train on a small labeled set, then label and retrain on the unlabeled data.	Simple and scalable; good performance with confident classifiers.	Can propagate errors with incorrect pseudo-labels.	Medical imaging, speech recognition, web content classification	Chen et al. (2022)
Co-training	Two classifiers trained on different views label unlabeled data for each other.	Robust to noise; utilizes diverse feature sets.	Dependent on quality and independence of feature sets.	Sentiment analysis, document classification	Peng et al. (2020)
Graph-based	Use a graph of data similarity to propagate labels.	Effective in capturing data structure; good for high-dimensional data.	Poor scalability with large datasets.	Social networks, image classification	Song et al. (2022)

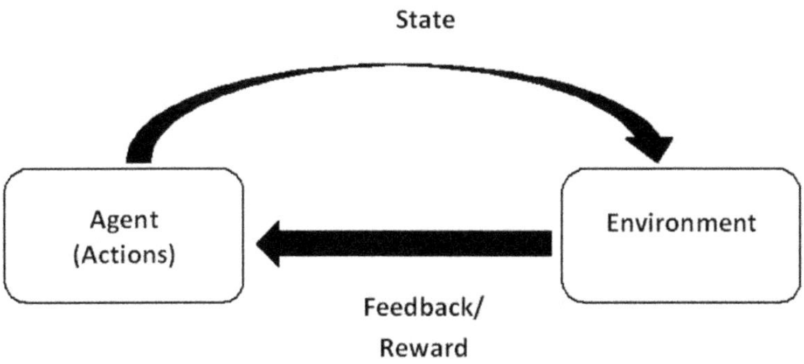

FIGURE 3.5 Reinforcement Learning

TABLE 3.4
Reinforcement Learning Algorithms

Method	Description	Advantages	Limitations	Applications
Q-Learning	A value-based approach using a Q-table to iteratively learn the best actions.	- Guaranteed convergence - Simple and interpretable	- No generalization - Scalability issues	- Robotics - Games - Autonomous Vehicles - Portfolio Management
Deep Q-Network (DQN)	Uses deep neural networks to approximate Q-values for high-dimensional states.	- Handles complex, high-dimensional input spaces	- Inefficient in sparse-reward environments - Struggles with exploration	- Video Games - Robotics - Simulation Environments
Policy Gradient Methods	Directly optimize the policy to maximize expected reward.	- Effective in continuous or large action spaces	- Prone to high variance - Can be inefficient without advanced optimization techniques	- Robotics control - Continuous control tasks

example, teaching a child to walk, where it gets rewards for maintaining balance and moving forward.

The algorithms for RL are presented in Table 3.4. The Q-learning algorithm learns the optimal action-value function $Q(s,a)$. This function represents the expected reward on taking action a in state s. In Deep Q Networks (DQN) algorithm, deep neural networks are used to approximate Q-values for high-dimensional environments.

Various policy gradient methods such as Proximal Policy Optimization (PPO), Advantage Actor-Critic (A3C), and Deep Deterministic Policy Gradient (DDPG) are used in various applications of RL. The PPO method encourages exploration and exploitation effectively, A3C employs multiple agents to accelerate training, and DDPG suits action spaces with continuous actions (Shakya et al., 2023).

RL is extensively used in robotics, where robots perform the tasks by trial and error. It has a wide range of applications in game playing in AI systems such as AlphaGo and OpenAI Gym for game mastering. RL facilitates self-driving cars to make choices within changing environments.

3.6 SELF-SUPERVISED LEARNING

The self-supervised learning is a form of ML in which models learn from unlabeled data by creating their own supervisory signals (Wang et al., 2022). In contrast to supervised learning, which needs human-labeled data, self-supervised learning generates labels from the data itself using a pretext task, allowing models to learn useful representations without direct human annotation. The self-supervised learning involves mainly two phases:

- *Pretext Task (Pertaining Phase)*: The model creates pseudo-labels out of the data and learns to represent by figuring out a designed task (such as predicting parts of an image or sentence missing).
- *Fine-Tuning (Downstream Task)*: The model, after pretraining, is then fine-tuned on a reduced labeled dataset for a particular task, such as classification or object detection.

Some of the key algorithms of Self supervised learning are mentioned in Table 3.5. Figure 3.6 shows the process of performing self-supervised learning.

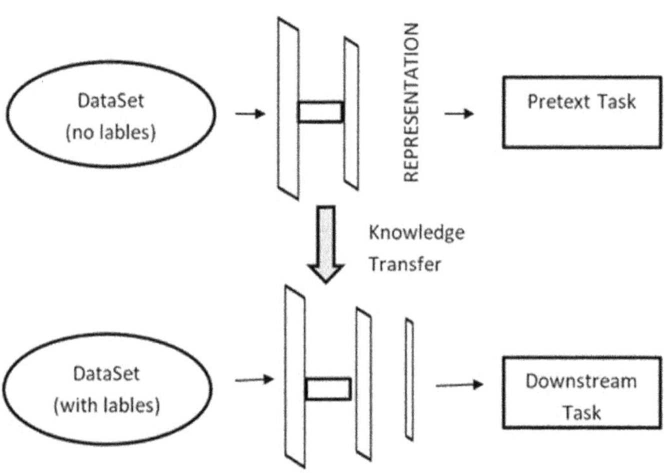

FIGURE 3.6 Self-Supervised Learning

TABLE 3.5
Self-Supervised Learning Algorithms

Method	Advantages	Limitations	Applications
Contrastive Learning	- Excellent performance in low-label or zero-shot settings - Effective for representation learning	- Sensitive to batch size - Requires careful selection of negative samples	NLP, image representation, audio and video processing
Transformer-Based Models	- No need for manual annotation (uses pretext tasks) - Captures long-range dependencies with self-attention	- High computational cost - Requires large volumes of training data	NLP, vision-language tasks, code understanding

3.6.1 Algorithms for Self-Supervised Learning

- *Contrastive Learning*: Contrastive learning learns representations by contrasting similar and dissimilar points in data. It brings similar instances (positive pairs) closer and pushes different instances (negative pairs) apart. It includes models such as SimCLR (Simple Contrastive Learning Representation) and MoCo (Momentum Contrast) (Chen et al., 2020). The SimCLR applies data augmentation (cropping, color jittering, etc.) to produce various views of the image. A neural network is used to bring closer representations of the same image and push away different images' representations. MoCo employs a momentum encoder to dynamically build a big dictionary of negative samples. It aids learning efficiently by effectively managing large-scale data.
- *Transformer-Based Models*: The Transformer model is well-suited for modeling long-range dependencies efficiently and thus suitable for learning general-purpose representations from raw unlabeled data (Islam et al., 2024). In self-supervised learning, the models are trained on automatically generated labels (pretext tasks) that need not be annotated manually. The Transformer acquires useful representations by learning these pretext tasks and can later be fine-tuned to use them for downstream tasks. Transformer-based models are extensively used in NLP, computer vision and multimodal tasks. BERT (Bidirectional Encoder Representations from Transformers) is one of the Transformer-based neural network architectures. It is pretrained for the prediction of missing words in sentences (masked language modeling) using self-supervised learning.

It is fine-tuned later for sentiment analysis, text summarization, and question-answering. Self-supervised models such as SimCLR learn strong image features in computer vision without labels.

These learned representations are then employed for downstream applications such as image classification and object detection with little labeled data. Self-supervised learning is employed to train speech recognition, speaker identification, and music classification models. In video analysis, models learn representations from unlabeled videos by predicting missing frames or temporal relationships. Transformer models have transformed self-supervised learning, especially in NLP and computer vision. It is employed widely in NLP applications such as text classification and translation. The BERT Transformer model learns deep contextualized representations by pre-training masked words in sentences (Masked Language Modeling); GPT (Generative Pre trained Transformer) utilizes autoregressive pretraining, through which it forecasts the subsequent word in a sequence. It empowers sophisticated language models such as ChatGPT for text generation and comprehension.

3.7 GENERATIVE LEARNING

Generative learning model learns by extracting meaningful associations and relations between existing knowledge, information, and experiences. It learns and creates new text, images, audio, executable computer programs, mathematical equations, and ideas. GenAI models use different learning paradigms, primarily unsupervised learning, self-supervised learning, and RL. Generative learning is the subfield of ML in which models learn to generate new data instances that are similar to a dataset. Unlike discriminative models, which classify points, generative models intend to gain understanding of the data distribution and generate samples from it. Generative learning algorithms have become a powerful tool in various application fields in AI because they can represent complex data distributions, create realistic content, and learn from sparse or unlabeled data. This allows generative learning to be applied broadly in applications like image synthesis, text generation, drug discovery, and anomaly detection. It is achieved using various methodologies, such as probabilistic modeling, DL architectures, and statistical techniques. Table 3.6 summarizes the different learning algorithms used for generative learning.

The generative learning algorithms are classified into three types: unsupervised, self-supervised, and reinforcement learning.

In unsupervised generative learning, generative models learn to approximate the empirical probability distribution, $p(x)$. They generate novel samples by directly approximating data distribution. Most of GenAI models learn from unlabeled data, discovering hidden patterns in datasets. It can be accomplished through explicit and implicit density models. The explicit defines a target probability distribution p and learns parameters for it, such as for VAEs learn probabilistic representations of data to generate new, similar data samples, while implicit density models learn to generate samples without specifying the distribution, for example, GANs. The different evaluation metrics for generative learning are as follows:

- *Inception Score (IS)*: Measures sample quality.
- *Frechet Inception Distance (FID)*: Measures the similarity between real and generated distributions.
- *Perceptual Path Length (PPL)*: Evaluates latent space smoothness.

Unsupervised generative learning is a revolutionary approach that allows machines to model complex data distributions without requiring labeled data.

TABLE 3.6
Generative Learning Algorithms

Type	Description	Advantages	Limitations	Applications
Unsupervised Generative Learning	Learns patterns and data distributions without using labeled datasets.	- No need for labeled data - Suitable for vast unlabeled datasets	- Training instability - Requires fine-tuning	- Image generation - NLP - Biomedical imaging - Synthetic data - Gaming
Self-Supervised Generative Learning	Generates supervisory signals from raw data using pretext tasks.	- Task-agnostic learning - Scales well with data	- Performance depends on pretext task design - May need domain-specific tuning	- Language modeling - Representation learning - Vision tasks
Generative Reinforcement Learning (GRL)	Combines generative models with reinforcement learning to simulate environments or generate policies.	- Simulates experience to improve sample efficiency - Helps in complex environments	- Risk of generating unrealistic/ hallucinated experiences	- Robotics - Game AI - Simulation training

With advancements in DL, it continues to play a crucial role in fields like computer vision, NLP, healthcare, and cybersecurity for data synthesis, representation learning, anomaly detection, and feature extraction. Prominent examples include VAEs, GANs, normalizing flows, and diffusion models. The research trends on unsupervised generative learning focuses on improving model efficiency, interpretability, and robustness for real-world applications.

The self-supervised generative learning (SSGL) is a cutting-edge paradigm in AI that leverages self-supervised learning principles to train generative models. GenAI models, like GPT (Generative Pre-trained Transformer), use self-supervised learning, in which they predict missing parts of the data. For example, the GPT models are trained by predicting the next word in a sequence without requiring human-labeled data.

Generative reinforcement learning (GRL) refers to a class of RL approaches that integrate generative models such as GANs, VAEs, and autoregressive models to simulate environment dynamics, generate synthetic experience, or shape reward signals. Some GenAI models use Reinforcement Learning with Human Feedback (RLHF) to fine-tune outputs based on rewards. For example, ChatGPT uses RLHF to make responses more human-like and contextually appropriate.

3.8 SUMMARIZATION AND COMPARISON OF DIFFERENT LEARNING PARADIGMS

Table 3.7 lists the summarization of various learning algorithms discussed so far in this chapter.

TABLE 3.7
Summarization and Comparison of Different Learning Paradigms

Learning Paradigm	Type of Data Required	Objective	Typical Algorithms/Models	Advantages	Limitations
Supervised Learning	Labeled data (input-output pairs)	Learn a mapping from inputs to known outputs; minimize prediction error	Linear Regression, Logistic Regression, Decision Trees, SVM, Neural Networks	High accuracy with sufficient labeled data; straightforward evaluation	Requires large labeled datasets; prone to overfitting
Unsupervised Learning	Unlabeled data	Discover hidden patterns, structures, or relationships within data	K-Means, Hierarchical Clustering, PCA, Autoencoders	No need for labeled data; good for exploratory analysis	Difficult to evaluate; patterns may be ambiguous
Semi-Supervised Learning	Mix of labeled and unlabeled data	Utilize limited labeled data with large unlabeled data to improve learning	Label Propagation, Semi-Supervised SVM, Graph-based Learning, Pseudo-labeling	Reduces labeling cost; better generalization than supervised on small labeled sets	Performance depends on quality of unlabeled data; complex training
Reinforcement Learning	Interaction data (states, actions, rewards)	Learn optimal actions through trial and error to maximize cumulative reward	Q-Learning, Deep Q-Network (DQN), Policy Gradient, Actor-Critic	Effective for sequential decision problems; learns from environment	Requires extensive training; instability in complex environments
Self-Supervised Learning	Unlabeled data with internally generated pseudo-labels or pretext tasks	Learn useful representations without external labels	Contrastive Learning (SimCLR), Masked Language Modeling (BERT), MAE, BYOL	Reduces need for labeled data; strong feature learning; improves transfer learning	Designing good pretext tasks is non-trivial; computationally intensive
Generative Learning	Labeled or unlabeled data depending on model	Generate new data samples similar to training distribution	GANs, VAEs, Diffusion Models, Normalizing Flows	Can create realistic data; useful for data augmentation and creativity	Training instability; risk of mode collapse; evaluation is difficult

3.9 EMERGING AND HYBRID LEARNING PARADIGMS

The applications of applied GenAI are much more advanced than those of traditional supervised and unsupervised learning methods. Modern researchers build hybrid learning frameworks that combine multiple learning methods to encourage flexibility, efficiency, and generalization. These paradigms aim to make AI systems more data-efficient, context-sensitive, and able to process complex real-world tasks involving heterogeneous data types and continuous learning needs. This section discusses some of the recent emerging and hybrid learning paradigms. These are as follows:

- Transfer learning and fine-tuning
- Federated learning
- Active learning
- Meta-learning (learning to learn)
- Continual and lifelong learning
- Multimodal learning (text + image + audio)
- Neuro-symbolic and hybrid models

3.9.1 Transfer Learning and Fine-Tuning

Transfer learning allows a model trained on one task or domain to be applied to another related task. Instead of learning from the ground up, a pretrained model (such as GPT, BERT, or ResNet) transfers learned representations to another task with minimal data (Vrbančič & Podgorelec, 2020)

Fine-tuning follows and fine-tunes the model by updating only specific layers or parameters on the target data, producing faster training and better performance.

This approach lies at the heart of GenAI, rendering big models trained on vast corpora domain-adjustable toward uses in healthcare, law, or education.

3.9.2 Federated Learning

Federated learning is a distributed approach where heterogeneous devices or nodes simultaneously learn a collective model without exposing raw data. Each client learns independently and sends model updates to a central server for aggregation (Mammen, 2021).

This ensures data privacy and security, and hence it is suitable for critical applications like finance and healthcare. It also reduces data transfer overhead and promotes cooperative intelligence in distributed systems.

3.9.3 Active Learning

With active learning, the model actively queries the most informative or confusing samples from an unlabeled dataset to be labeled by a specialist. This prevents high labeling costs and helps achieve maximum learning efficiency (Settles, 2009).

For GenAI, active learning allows models to learn more efficiently with less training, especially for niche fields where data is expensive to obtain or not readily available.

3.9.4 Meta-Learning (Learning to Learn)

Meta-learning refers to learning models that can learn new tasks efficiently with minimal data—typically described as "learning how to learn" (Zhou et al., 2018).

Instead of becoming proficient at one task, a meta-learner acquires transferable knowledge across tasks and can adapt quickly.

In GenAI, meta-learning enables adaptive and personalized AI systems that are able to fine-tune for a particular user or new environment with minimal retraining.

3.9.5 Continual and Lifelong Learning

Continual learning (or lifelong learning) enables models to build up new information over time without forgetting what was learned previously—a problem known as "catastrophic forgetting."

Such systems mimic human learning, with increasing understanding and the capacity to adapt to new data (Parisi et al., 2019).

In GenAI, continuous learning allows models to learn and evolve with changing data patterns and user behavior, for ongoing performance and relevance.

3.9.6 Multimodal Learning (Text + Image + Audio)

Multimodal learning integrates data from multiple sources or sensory modalities—text, image, video, and audio—to create more informative, context-aware models.

By integrating multiple modalities, these models can generate more realistic and coherent output (e.g., text-to-image or text-to-video generation) (Li et al., 2017).

3.9.7 Neuro-Symbolic and Hybrid Models

Neuro-symbolic models combine the pattern recognition power of neural networks with the "logical reasoning capability of symbolic AI."

In GenAI, neuro-symbolic systems enhance interpretability, reasoning, and reliability (Lappin, 2025).

3.10 CASE STUDIES

3.10.1 Supervised Learning Case Study—Diagnosis of Disease in Healthcare

Domain: Healthcare
Goal: Automatically diagnose diabetic retinopathy from retinal fundus images.
Type of Data: Labeled images (eye disease stage labels by ophthalmologists).

Approach:

Train a CNN on labeled datasets like EyePACS or APTOS 2019. Each image is labeled as No DR, Mild, Moderate, Severe, or Proliferative DR. The model is trained to learn the relationship between image features and disease classes (Gulshan et al., 2016).

Result:

The model trained on these data attains high diagnostic accuracy comparable to specialist ophthalmologists, enabling automatic screening in rural health centers.

Advantages:

- Lowers human diagnosis effort.
- Enables early detection and treatment.

Limitations:

- Has a tendency to require large labeled data sets and expert labeling.

3.10.2 Unsupervised Learning Case Study—Student Clustering in Education

Domain: Education

Objective: Identify learning patterns and student clusters for tailored instruction.

Type of Data: Unlabeled student performance data (scores, activity logs, engagement measures).

Method:

K-Means Clustering classifies students according to their learning behavior patterns extracted from Learning Management Systems (LMS) (Tuyishimire et al., 2022).

Cluster 1: High achievers, Cluster 2: Persistent learners, Cluster 3: Challenged students.

Outcome:

Teachers gain insights to adjust teaching interventions, provide additional assistance, and design personalized assignments.

Advantages:

- Uncover deep-seated learning patterns.
- No labeling effort involved.

Disadvantages:

- Meaning of clusters may require domain knowledge.

3.10.3 Semi-Supervised Learning Case Study—Medical Image Segmentation

Domain: Healthcare

Goal: Segment tumors or organs in MRI scans with limited labeled data.

Data Type: Few labeled MRI scans + large number of unlabeled scans.

Method:

A U-Net model is trained on a mixture of labeled and unlabeled images by applying a consistency regularization method. The model utilizes unlabeled images to discover spatial patterns and enhance segmentation edges (Jiang et al., 2023).

Result:

Correct segmentation with fewer labeled examples, minimizing reliance on costly manual annotation by radiologists.

Benefits:

- Lowers cost of labeling.
- Better generalization than fully supervised models.

Drawbacks:

- Noisy unlabeled data can be problematic.

3.10.4 Reinforcement Learning Case Study—Autonomous Vehicle Control

Domain: Autonomous Vehicles

Goal: Facilitate self-driving vehicles to make optimal driving choices in changing situations.

Type of Data: Sequential sensor data (camera, LiDAR, GPS) with feedback (reward for safe routing).

Approach:

A Deep Q-Network (DQN) is trained by interacting with a simulated driving world. It receives rewards for remaining in lane, preventing collisions, and adhering to traffic lights (Kim et al., 2022).

Outcome:

The car acquires policies to steer, accelerate, and brake autonomously without being programmed.

Advantages:

- Learns the best strategies through exploration.
- Copes with difficult sequential decision-making.

Limitations:

- Takes massive training in simulation.
- Unsafe to train directly in actual settings.

3.10.5 Case Study on Self-Supervised Learning—Medical Report Generation

Domain: Healthcare

Objective: Produce diagnostic reports from medical images (e.g., chest X-rays).

Type of Data: Unlabeled X-ray images with sparse text reports.

Method:

A Vision Transformer (ViT) is pretrained via self-supervised contrastive learning (SimCLR or MAE) to learn image representations. Subsequently, it is fine-tuned on a small labeled set that maps images to text diagnoses (Rani et al., 2024).

Outcome:

Better report generation accuracy and interpretability with sparse labeled data.

Advantages:

- Leverages large amounts of unlabeled medical data.
- Improves downstream diagnosis performance.

Limitations:

- Needs careful pretext task design.
- Has extremely high computational cost for pretraining.

EXERCISES

Multiple Choice Questions (MCQs)

1. Which of the following best describes a learning paradigm in AI?
   a) A mathematical function
   b) One means of data storage
   c) A framework that defines how a model learns from data
   d) Data visualization technique
2. The three main kinds of AI learning paradigms are:
   a) Supervised, unsupervised, and reinforcement learning
   b) Structured, semi-structured, and unstructured learning
   c) Static, dynamic, and adaptive learning
   d) None of these
3. The aim of any AI learning paradigm is to:
   a) Reduce training time
   b) Increase data redundancy
   c) Permit machines to self-improve with experience
   d) Decrease the model's accuracy
4. Which one of the following algorithms is used in supervised learning?
   a) K-Means
   b) Decision trees
   c) PCA
   d) Autoencoders
5. The primary objective of clustering is to:
   a) Minimize reward
   b) Group similar data points together
   c) Predicting future values
   d) Label known outputs
6. Semi-supervised learning uses:
   a) Only labeled data
   b) Only unlabeled data
   c) Labeled and unlabeled data combined
   d) Reinforcement rewards

7. Which of the following combines characteristics of both supervised and unsupervised learning?
   a) RL
   b) SSL
   c) Transfer learning
   d) Self-supervised learning
8. In RL, the feedback is given in the form of:
   a) Correct answers
   b) Reward or penalty
   c) Class labels
   d) Hidden states
9. Self-supervised learning finds prominent applications in:
   a) Automatic labeling of unlabeled data
   b) Reinforcement signal optimization
   c) Maximization of direct reward
   d) Random data generation
10. Generative learning focuses on:
   a) Generating new data samples similar to training data
   b) Classifying input data
   c) Reinforcing agent actions
   d) Reducing model complexity

Short/Long Answer Questions

1. Define the term "AI learning paradigm."
2. Enumerate the major categories of learning paradigms in AI.
3. Distinguish between classification and regression problems.
4. What is self-supervised learning?
5. What are GANs, and how do they work?
6. Name two applications of generative learning.
7. Explain the workflow of a supervised learning model using a real-world example.
8. How is self-supervised learning different from supervised learning?
9. Name two common applications of self-supervised learning.
10. Explain the process of self-supervised learning with a suitable example.
11. Discuss the role of self-supervised learning in modern AI systems such as GPT or BERT models.
12. Explain the architecture and working of a GAN.
13. Compare discriminative and generative models using examples.
14. Discuss the importance of generative learning in the era of GenAI.
15. Distinguish between model-based and model-free RL.
16. Discuss real-world applications, such as robotics and gaming, which make use of RL.

REFERENCES

Breiman, L. (2001). Random forests. *Machine learning*, *45*, 5–32.

Chen, B., Jiang, J., Wang, X., Wan, P., Wang, J., & Long, M. (2022). Debiased self-training for semi-supervised learning. *Advances in Neural Information Processing Systems, 35*, 32424–32437.

Chen, T., He, T., Benesty, M., Khotilovich, V., Tang, Y., Cho, H., . . . & Zhou, T. (2015). Xgboost: Extreme gradient boosting. *R Package Version 0.4–2, 1*(4), 1–4.

Chen, T., Kornblith, S., Norouzi, M., & Hinton, G. (2020, November). A simple framework for contrastive learning of visual representations. In *International conference on machine learning* (pp. 1597–1607). PmLR.

Gulshan, V., Peng, L., Coram, M., Stumpe, M. C., Wu, D., Narayanaswamy, A., . . . & Webster, D. R. (2016). Development and validation of a deep learning algorithm for detection of diabetic retinopathy in retinal fundus photographs. *Jama, 316*(22), 2402–2410.

Guo, G., Wang, H., Bell, D., Bi, Y., & Greer, K. (2003). KNN model-based approach in classification. In *On the move to meaningful internet systems 2003: CoopIS, DOA, and ODBASE: OTM confederated international conferences, CoopIS, DOA, and ODBASE 2003, Catania, Sicily, Italy, November 3–7, 2003. Proceedings* (pp. 986–996). Berlin, Heidelberg: Springer.

Islam, S., Elmekki, H., Elsebai, A., Bentahar, J., Drawel, N., Rjoub, G., & Pedrycz, W. (2024). A comprehensive survey on applications of transformers for deep learning tasks. *Expert Systems with Applications, 241*, 122666.

James, G., Witten, D., Hastie, T., Tibshirani, R., & Taylor, J. (2023). Linear regression. In *An introduction to statistical learning: With applications in python* (pp. 69–134). Cham: Springer International Publishing.

Jiang, Y., Sui, X., Ding, Y., Xiao, W., Zheng, Y., & Zhang, Y. (2023). A semi-supervised learning approach with consistency regularization for tumor histopathological images analysis. *Frontiers in Oncology, 12*, 1044026.

Kim, M. S., Eoh, G., & Park, T. H. (2022). Decision making for self-driving vehicles in unexpected environments using efficient reinforcement learning methods. *Electronics, 11*(11), 1685.

Lappin, S. (2025). Neuro-symbolic models of AI. *Journal of Applied Logics, 12*(6), 1631–1642.

LaValley, M. P. (2008). Logistic regression. *Circulation, 117*(18), 2395–2399.

Li, H., Zhu, J., Ma, C., Zhang, J., & Zong, C. (2017, September). Multi-modal summarization for asynchronous collection of text, image, audio and video. In *Proceedings of the 2017 conference on empirical methods in natural language processing* (pp. 1092–1102). Association for Computational Linguistics.

Mahadevkar, S. V., Khemani, B., Patil, S., Kotecha, K., Vora, D. R., Abraham, A., & Gabralla, L. A. (2022). A review on machine learning styles in computer vision—techniques and future directions. *Ieee Access, 10*, 107293–107329.

Mammen, P. M. (2021). Federated learning: Opportunities and challenges. *arXiv preprint arXiv:2101.05428*.

Mazyavkina, N., Moustafa, S., Trofimov, I., & Burnaev, E. (2021, July). Optimizing the neural architecture of reinforcement learning agents. In *Intelligent computing: Proceedings of the 2021 computing conference, volume 2* (pp. 591–606). Cham: Springer International Publishing.

McDonald, G. C. (2009). Ridge regression. *Wiley Interdisciplinary Reviews: Computational Statistics, 1*(1), 93–100.

Parisi, G. I., Kemker, R., Part, J. L., Kanan, C., & Wermter, S. (2019). Continual lifelong learning with neural networks: A review. *Neural Networks, 113*, 54–71.

Peng, J., Estrada, G., Pedersoli, M., & Desrosiers, C. (2020). Deep co-training for semi-supervised image segmentation. *Pattern Recognition, 107*, 107269.

Pisner, D. A., & Schnyer, D. M. (2020). Support vector machine. In *Machine learning* (pp. 101–121). Academic Press.

Quinlan, J. R. (1996). Learning decision tree classifiers. *ACM Computing Surveys (CSUR), 28*(1), 71–72.

Rani, V., Kumar, M., Gupta, A., Sachdeva, M., Mittal, A., & Kumar, K. (2024). Self-supervised learning for medical image analysis: A comprehensive review. *Evolving Systems, 15*(4), 1607–1633.

Ranstam, J., & Cook, J. A. (2018). LASSO regression. *Journal of British Surgery, 105*(10), 1348–1348.

Rigatti, S. J. (2017). Random forest. *Journal of Insurance Medicine, 47*(1), 31–39.

Sarker, I. H. (2021). Machine learning: Algorithms, real-world applications and research directions. *SN Computer Science, 2*(3), 160.

Settles, B. (2009). Active learning literature survey. Computer Sciences Technical Report 1648, University of Wisconsin-Madison.

Shakya, A. K., Pillai, G., & Chakrabarty, S. (2023). Reinforcement learning algorithms: A brief survey. *Expert Systems with Applications, 231*, 120495.

Song, Z., Yang, X., Xu, Z., & King, I. (2022). Graph-based semi-supervised learning: A comprehensive review. *IEEE Transactions on Neural Networks and Learning Systems, 34*(11), 8174–8194.

Stoltzfus, J. C. (2011). Logistic regression: A brief primer. *Academic Emergency Medicine, 18*(10), 1099–1104.

Suthaharan, S. (2016). Support vector machine. In *Machine learning models and algorithms for big data classification: Thinking with examples for effective learning* (pp. 207–235). Boston, MA: Springer US.

Taye, M. M. (2023). Understanding of machine learning with deep learning: Architectures, workflow, applications and future directions. *Computers, 12*(5), 91.

Tuyishimire, E., Mabuto, W., Gatabazi, P., & Bayisingize, S. (2022). Detecting learning patterns in tertiary education using K-means clustering. *Information, 13*(2), 94.

Vrbančič, G., & Podgorelec, V. (2020). Transfer learning with adaptive fine-tuning. *IEEE Access, 8*, 196197–196211.

Wang, Y., Song, D., Wang, W., Rao, S., Wang, X., & Wang, M. (2022). Self-supervised learning and semi-supervised learning for multi-sequence medical image classification. *Neurocomputing, 513*, 383–394.

Webb, G. I., Keogh, E., & Miikkulainen, R. (2010). Naïve Bayes. *Encyclopedia of Machine Learning, 15*(1), 713–714.

Yang, F. J. (2018, December). An implementation of naive bayes classifier. In *2018 International conference on computational science and computational intelligence (CSCI)* (pp. 301–306). IEEE.

Ying, L. U. (2015). Decision tree methods: Applications for classification and prediction. *Shanghai Archives of Psychiatry, 27*(2), 130.

Zhang, S. (2021). Challenges in KNN classification. *IEEE Transactions on Knowledge and Data Engineering, 34*(10), 4663–4675.

Zhou, F., Wu, B., & Li, Z. (2018). Deep meta-learning: Learning to learn in the concept space. *arXiv preprint arXiv:1802.03596.*

4 GenAI
Models and Architecture

4.1 GENAI FIVE-LAYER FRAMEWORK ARCHITECTURE

Gen AI models have played a crucial role in advancement of AI technologies in different application areas such as chatbots, image synthesis tools, and music generation systems. This chapter explores different models and architecture framework employed in development and implementation of GenAI.

The GenAI five-layered architecture provides a systematic view of how generative systems are built, deployed, and used. It can be organized into five essential architectural layers. Each layer plays a distinctive role but all layers work together to achieve an efficient AI system.

4.1.1 Data Processing Layer

This foundational layer focuses on data acquisition, storage, and preparation. The major tasks performed by this layer are as follows:

- This layer acquires data from diverse sources.
- It performs data cleaning and prepares it for its model analysis. This step ensures quality and consistency.
- It performs feature engineering to extract relevant signals, enhancing model training by minimizing noise.

4.1.2 Generative Model Layer

At this layer, AI models are developed and customized. It involves the following:

- Selects generative models such as LLMs for text or GANs for images, depending on the application domain.
- Implements model training or fine-tuning strategies tailored to specific tasks.
- Leverages transfer learning and domain adaptation to build on pretrained models and optimize for targeted performance.

4.1.3 Feedback and Continuous Learning Layer

This layer enables adaptive learning through real-world feedback. The key tasks of this layer are as follows:

- Gathers insights via user interactions, satisfaction surveys, and performance metrics.

DOI: 10.1201/9781003713951-4

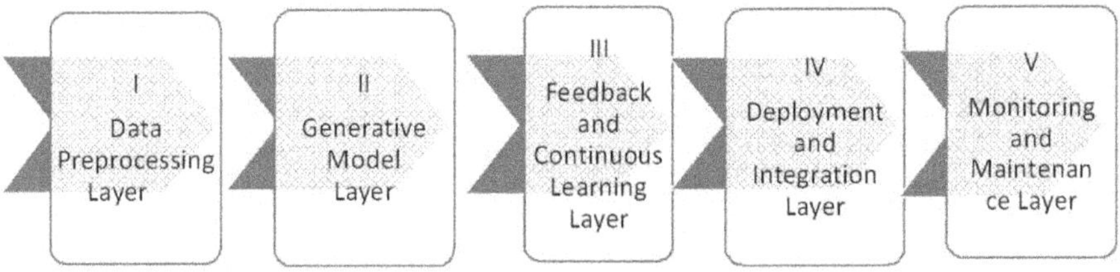

FIGURE 4.1 GenAI Five-Layered Architecture

- Applies RL and feedback mechanisms to iteratively improve model predictions.
- Identifies and mitigates issues such as bias, hallucination, or factual inaccuracies through structured feedback loops.

4.1.4 Deployment and Integration Layer

This operational layer focuses on real-world integration. The major focus areas of this layer are as follows:

- Deploys models through application programming interfaces (APIs), cloud-native platforms, or enterprise applications.
- Ensures seamless integration across front-end interfaces and back-end systems.
- Addresses key infrastructure challenges such as compute resource management, scalability, latency, and cybersecurity.

4.1.5 Monitoring and Maintenance Layer

This topmost layer ensures long-term sustainability and governance. The main functions of this layer are as follows:

- Continuously monitors model performance, reliability, and ethical compliance.
- Facilitates periodic retraining and updates based on new data or shifting business goals.
- Focuses on cost management, version control, and risk mitigation to ensure consistent AI service quality.

Figure 4.1 illustrates a **vertical stack of five interdependent layers**, each representing a stage in the generative AI pipeline:

- *Data Processing Layer (Base Layer)*: The input layer that prepares and feeds high-quality data into the system.
- *Generative Model Layer*: Builds and fine-tunes the AI models based on the curated data.

- *Feedback and Continuous Learning Layer*: Acts as a loop around the model, improving its performance through dynamic feedback.
- *Deployment and Integration Layer*: Connects the trained models with production systems and user applications.
- *Monitoring and Maintenance Layer (Top Layer)*: Ensures the system stays reliable, ethical, and efficient over time.

Each layer feeds into the next while also maintaining feedback connectivity, ensuring an adaptive and sustainable AI ecosystem.

4.2 GENAI ARCHITECTURE MODELS

There are several generative models with different mechanisms and uses. These are discussed in the following.

4.2.1 Generative Adversarial Networks

GANs are made up of two adversarial neural networks: Generator (G) and Discriminator (D).

The Generator network creates artificial data by mapping random noise to meaningful outputs, whereas the discriminator network determines whether the input data provided is real (from the dataset) or artificial (produced by G). These networks play a zero-sum game in which the Generator becomes better at deceiving the discriminator, and the discriminator becomes better at identifying fakes. With time, the Generator generates very realistic outputs. Therefore, the Generator improves until it creates outputs indistinguishable from real data (Goodfellow et al., 2020).

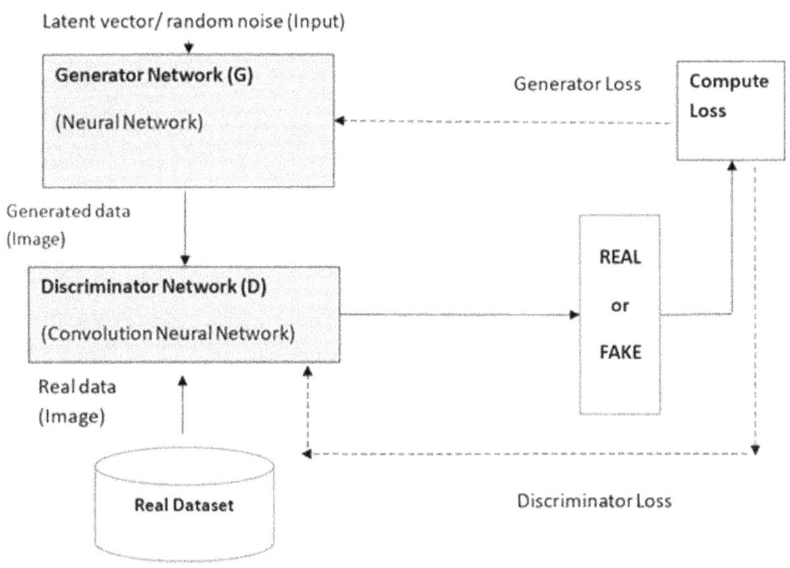

FIGURE 4.2 GAN Architecture

Figure 4.2 shows the architecture of a GAN model. The Generator accepts a random noise vector as input. The Generator network is a neural network that maps the noise to a structured output, for example, an image. The Generator produces a fake sample. The Discriminator is provided with real data (from the dataset) and fake data (produced by the generator). The Discriminator is a CNN that decides whether the input is real or fake. The Generator is trained based on how successfully it deceives the Discriminator, whereas the Discriminator is trained to be able to tell real and fake samples apart. Both Generator and Discriminator play adversely to each other. On receiving the Discriminator decision for "Fake" or "Real," the loss is computed using Equation (4.1):

$$G_{\min} D_{\max} V(D,G) = E_x[\log D(x)] + E_z\left[\log\left(1 - D\left(G(z)\right)\right)\right] \tag{4.1}$$

where

$G(z)$ is Generator output on receiving input noise z.
$D(x)$ is Discriminator probability that original data x is real.
$D(G(z))$ is the Discriminator probability that synthetic data $G(z)$ is real.
E_x is the Expected value over all original data instances.
E_z is the Expected value over all synthetic data instances.

As indicated in Equation (4.1), the Generator could not directly affect $[\log D(x)]$, so it tries to minimize the $\log\left(1 - D\left(G(z)\right)\right)$. On the contrary, during classification process, the Discriminator tries to achieve the correct label among "Fake" or "Real," so it tries to maximize $[\log D(x)]$.

The GAN has widespread applications. It is widely used for generating realistic images, synthesize short videos from learned patterns, generating fake images, audio and videos using various AI tools and editing software (Creswell et al., 2018).

4.2.2 Variational Autoencoders

VAE is a form of generative DL model that is composed of conventional autoencoders with probabilistic modeling to support variational inference (Pinheiro et al., 2021). This allows the model to create new data samples by capturing the underlying distribution of the input data. VAEs are applied extensively to areas like dimensionality reduction, data generation, latent space exploration, and game design. (Mak et al., 2023).

To understand the concept of VAEs, idea of latent space needs exploration. The term "latent space" refers to the collection of latent variables of an input image. It helps to know about the underlying distribution of data. The autoencoders model the latent space using dimensionality reduction. It compresses the data leaving only important and relevant features called latent space.

Many of the autoencoders are designed to produce discrete and fixed latent variable representation, but VAE represents the latent space as a continuous and probabilistic representation. It helps VAE to reconstruct the original input accurately. The neural network design of the VAE was first introduced in a 2013 paper by Diederik

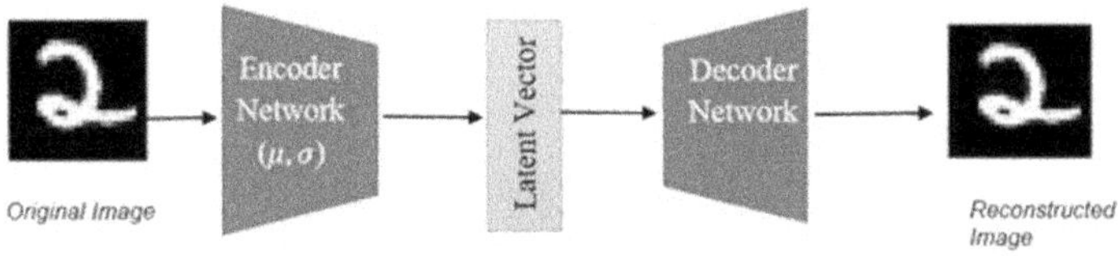

FIGURE 4.3 Block Diagram Variational Autoencoders

P. Kingma and Max Welling, entitled *Auto-Encoding Variational Bayes* (Kingma & Welling, 2013). The main goal of a VAE is to learn a low-dimensional, latent representation of the input data that preserves its underlying structure. This is performed by jointly training two neural networks: an encoder and a decoder.

As illustrated in Figure 4.3, the VAE architecture is constructed to balance good data compression with good reconstruction. It has two primary components:

- *Encoder*: It transforms the input data to a lower-dimensional latent space. It begins with an input layer of equal size to the features, followed by a series of decreasingly smaller hidden layers.

  The last layer of the encoder, the latent layer, predicts the parameters (mean μ and standard deviation σ) of a multivariate Gaussian distribution. This allows the generation of samples from the latent space, adding randomness to the learning process.
- *Decoder*: Reconstructs the original data from the sampled latent variables.

The probabilistic structure enables the VAE to produce diverse outputs and learn useful representations of complicated data. VAE maps input data to a lower-dimensional latent space and reconstruct it with minor variations. VAEs differ from standard autoencoders, in that they add probabilistic latent variables, enabling them to learn a smooth data distribution.

4.2.2.1 Variational Autoencoder Loss Function

The loss function of VAE is computed on the basis of a reconstruction loss and divergence loss. The reconstruction loss is measured by calculating the difference between predicted probability distribution and actual data distribution of dataset. It is represented in Equation (4.2) as:

$$L\left(X, X'\right) \tag{4.2}$$

The reconstruction loss is quite helpful in reconstructing the input images; however, it does not learn improved mean (μ) and standard deviation (σ) (Odaibo, 2019).

The K–L divergence or KL divergence loss helps to create better μ and σ that further creates not effective latent space. It is denoted by D KL(P||Q) and indicates the divergence of one probability distribution over another. The KL divergence loss is represented in Equation (4.3) as:

$$KL(P\,||\,Q) = \sum P(X) * \log \frac{P(X)}{Q(X)} \tag{4.3}$$

In the above equation:

$P(X)$ and $Q(X)$ are the probability distribution on same variable X. Here, Q is the predicted distribution and P is the prior distribution.

As it is a loss, therefore minimization of KL divergence is always required. We can also maximize the equation given below as:

Therefore, total loss is given in Equation (4.4) as follows:

$$\text{Loss} = L(X, X') + \text{KL}(P\,||\,Q) \tag{4.4}$$

The VAE has a wide range of applications such as image synthesis for variations of images, for example, changing facial expressions or styles; anomaly detection to detect deviations from normal patterns, which is beneficial in cybersecurity and medical diagnostics; and drug discovery to assist in discovering molecular structures by creating new chemical compounds with good properties. VAE is also used for noise removal in dataset by reconstructing the data again from the latent space (Cui et al., 2025).

4.2.3 Transformer-Based Models

Transformers have shown success in NLP domain and have become a major breakthrough in computer vision. Transformer models make use of self-attention techniques in order to process sequences of information in an efficient manner. In contrast to the traditional RNNs, Transformers process long-range dependencies better and allow for parallel processing (Gillioz et al., 2020).

After their success in NLP, the Transformers are used for application in visual data, resulting in the Vision Transformers (ViTs). The models process images as sequences of patches and use the same self-attention mechanism to learn strong visual representations. Several studies have established that Transformers are effective in many vision tasks, from image classification to object detection and segmentation.

The strong performance of Transformers in the vision space has prompted a move toward incorporating them into generative models. Many researches are performed on substituting the standard CNN model employed in GANs with Transformer-based models. With the potential of self-attention, Transformer-based GANs have the ability to generate high-quality, coherent images with enhanced long-range structural consistency. This change has unlocked new directions in GenAI, expanding the limits of image generation and innovation.

4.2.3.1 Examples and Applications

Table 4.1 shows different models used for different domains, with their description and applications. Some of the popular Transformer models include GPT model that is applied for text generation, conversational AI, and creative writing; BERT that

TABLE 4.1
Examples and Applications of Generative Transformer Models

Domain	Model	Description	Applications
Text Generation	GPT-3/GPT-4	LLMs trained on vast text data for various NLP tasks	Chatbots, content creation, code generation, summarization
	T5	Unified text-to-text model for multiple NLP tasks	Translation, summarization, text classification
Image Generation	DALL·E/DALL·E 2/ DALL·E 3	Text-to-image models generating realistic visuals from descriptions	Design automation, concept art, marketing content
	ViT-GAN	GAN with a ViT-based generator	High-resolution image synthesis, synthetic face generation
	TransGAN	Fully Transformer-based GAN architecture	Unsupervised image generation, style transfer, texture synthesis
Video Generation	VideoGPT/ Transformer	Transformer models predicting future video frames	Video editing, synthetic video creation, game prototyping
Audio Generation	Jukebox (OpenAI)	Music generation model using Transformers	AI-composed music, soundtrack generation, creative composition
	AudioLM	Coherent long-form speech/ audio generation	Voice synthesis, podcast narration, audio storytelling
Multimodal	Flamingo, GPT-4 (Multimodal)	Models processing text, image, and/or video inputs	Visual question answering, educational tools, interactive AI
	Imagen (Google)	Text-to-image model with diffusion and Transformer-like architecture	Scientific visualization, illustration, content creation

enhances language comprehension, commonly applied in search engines and sentiment analysis; and DALL·E which transforms text prompts into extremely detailed images (Mehmood et al., 2023).

4.2.4 Diffusion Models

Diffusion models are a type of generative model in ML that have gained attention for their ability to produce high-quality, realistic data, especially images, audio, and video. They are applied in tools such as DALL·E 2, Imagen, Midjourney, and Stable Diffusion. Diffusion models produce data by reversing a noise-injection process. Diffusion models learn data distribution by iteratively adding noise to genuine samples and then training a neural network to undo the process, gradually recovering the original data (Croitoru et al., 2023).

Diffusion models operate in two phases:

- *Forward Process (Diffusion)*: The forward process begins with an actual data. Then, we incrementally add noise to it repeatedly until the data becomes all noisy. This is a fixed, untrainable process.
- *Reverse Process (Denoising)*: The model works step by step to reverse the noise from the data. It begins with plain noise and then learns to output a clean image by gradually erasing noise. This reverse process is represented as a neural network. Therefore, the model learns to transition from "noise → image" by learning to remove noise gradually.

Table 4.2 compares different GenAI models.

TABLE 4.2
Comparison Table of Different GenAI Models

Aspect	GAN (Generative Adversarial Network)	VAEs (Variational Autoencoder)	Transformer Models (e.g., GPT, BERT)	Diffusion Models (e.g., Stable Diffusion, DALL·E 2)
Architecture	Generator + Discriminator (2 networks)	Encoder + Decoder + Latent space	Attention-based encoder–decoder or decoder	Denoising autoencoder-based iterative refinement
Training Type	Adversarial learning	Probabilistic + Reconstruction loss	Supervised/ Self-supervised	Noise-to-data via diffusion and reverse process
Latent Space	Implicit	Explicit and structured	No explicit latent space	Explicit noise-to-data latent mapping
Output Quality	High fidelity (photorealistic)	Smooth but blurry outputs	High quality (text, code, multimodal)	Extremely high fidelity (especially images)
Stability	Training is unstable, prone to collapse	Stable and easy to train	Stable but resource-heavy	Stable but slow to sample
Control and Interpretability	Difficult to control or interpret	Interpretable latent space	Controllable via prompts	Controllable via conditioning and prompts
Strengths	Sharp realistic images, fast generation	Good for interpolation, representation	Language generation, context handling	High-quality image generation, diverse samples
Weaknesses	Mode collapse, training instability	Blurry outputs, less sharpness	Requires large datasets and compute	Slow sampling, computationally intensive
Use Cases	Image generation, super-resolution, deepfakes	Data compression, anomaly detection	Text generation, translation, summarization	Text-to-image, inpainting, image editing

4.3 REAL-WORLD APPLICATIONS AND USE CASES OF GENAI

GenAI has moved rapidly from research labs into production systems. It has widespread applications of diverse areas such as in healthcare, entertainment, and finance (Madaan et al., 2024). This section presents detailed study of some of the applications of GenAI.

4.3.1 Healthcare

- *Medical Image Synthesis and Augmentation*: Generative models generate realistic medical images, including MRI, CT, and X-ray, to augment the training data for diagnostic models. In this way, model robustness is enhanced and overfitting is reduced, especially for rare conditions (Sai et al., 2024).
- *Clinical Note Generation and Summarization*: LLMs help clinicians by generating discharge summaries, progress notes, patient letters, among others, through structured data and voice transcripts, saving time for the clinicians themselves.
- *Drug Discovery and Molecular Design*: Generative models propose new molecules or suggest chemical modifications by sampling chemical space for early-stage drug candidate discovery.
- *Personalized Treatment Planning*: Synthetic patient simulations allow for the testing of scenarios regarding treatment strategies without exposing real patient data.

Key Benefit: Faster workflows, expanded datasets, and automatic documentation.
Risk Involved: The GenAI outputs are hallucinated or inaccurate. They may harm patients if not verified by humans.

4.3.2 Entertainment (Media, Games, Films)

It enables content creation and VFX. The tools generate concept art, textures, background scenery, and even full 3D asset placeholders (Sadiku et al., 2025). They accelerate storyboarding and visual effects.

- *Scriptwriting and Ideation of Stories*: LLMs suggest plot outlines, dialogues, and scene variations that enhance human creativity.
- *Synthetic Actors and Virtual Characters*: Voice cloning and face synthesis allow creators to prototype virtual performers or create nonhuman characters.
- *Personalized Experiences*: Games and interactive stories may be generated dynamically to match the player's preferences.
- *Music and Sound Design*: Models generate musical motifs, ambience, and adaptive soundtracks.

Key Benefits

- Drastic reduction of time and cost for creative iteration.
- Enabling smaller teams to deliver high-quality outputs.

Risks Involved

- Include copyright issues;
- Job displacement concerns; and
- Misuse of content, such as unauthorized use of a person's likeness.

4.3.3 Finance

- *Synthetic Data Creation*: GenAI models will generate realistic, privacy-preserving test datasets of trading algorithms, risk models, or fraud detectors when actual user data cannot be shared (Chelliah et al., 2025).
- *Automation and Summarization of Reports*: LLMs create draft investment memos, compliance summaries, and earnings highlights from raw financials.
- *Automation of Customer Service*: Chatbots and virtual assistants handle routine queries, onboarding, and document verification.

Key Benefits

- Faster model development cycles
- Improved privacy-preserving experimentation.

Primary Risk

- Model errors or bias cascading into financial decisions
- Lack of transparency in decision-making complicates regulatory compliance.

4.3.4 Notable Use Cases in Content Creation: Images, Video, Text

Images

- *Commercial Art and Advertising*: Quickly generate ad creative variations and A/B test concepts.
- *Stock Image Creation*: Royalty-free, on-demand images created according to a brief.
- *Photo Retouching and Restoration*: Remove artifacts, restore old photos, or turn sketches into photorealistic images.

Video

- *Automated Video Editing*: Auto-cut, scene transitions, and color grading suggestions.
- *Synthetic Video for Training*: Labeled video creation, like for surveillance or autonomous systems.

- *Deepfake Use in Controlled Creative Contexts*: Recreating a historical figure in a film with consent and legal clearance.

Text

- *Article and Copy Generation*: Creating blog posts, product descriptions, and marketing copy.
- *Summarization and Translation*: Condensing lengthy reports and translating them while preserving the tone.
- *Code Generation and Documentation*: Creating boilerplate code, unit tests, and API documents.

4.4 CONCLUSION

GenAI has rapidly emerged as one of the most transformative domains AI. In this chapter, we explored a range of foundational models and architectures that power modern generative systems, including probabilistic models and autoencoders through GANs, diffusion models, to large Transformer-based architectures. These models act as a backbone of generative revolution. Each of these brings unique strengths, thus allowing applications that span creative content generation, scientific discovery, data augmentation, simulation, and intelligent automation. In the future research, the generative system will surely turn out smarter, intelligent, and will align with human values and ethics.

EXERCISES

1. What are VAEs? How are they different from autoencoders?
2. Explain role of Transformers in generating languages on LLMs.
3. Explain the architecture of Generator and Discriminator in GAN model. How do they interact in data generation?
4. What are the different challenges faced by VAEs?
5. What is the significance of using diffusion models in generative tasks?
6. What is the KL divergence term in VAE, and why is it important?
7. Does Stable Diffusion approach achieve high-quality image generation? Give reasons.
8. Discuss the common problems in training GAN such as vanishing gradient.
9. Compare different generative models such as GANs, VAEs, Transformers, and diffusion models.
10. Which generative model is best suited for image, text, and music generation? Give reason in support of your answer.
11. How do GANs function, and what are their essential components?
12. What distinguishes GANs from VAEs and diffusion models?
13. What are the comparative strengths and weaknesses of diffusion models relative to GANs and VAEs?
14. In which scenarios are VAEs preferred over GANs for generative tasks?
15. How can the stability and quality of generative models be enhanced during their training?

16. What role do Transformer architectures play in modern GenAI?
17. How have Transformers revolutionized natural language generation and understanding?
18. How are Transformer-based models adapted for tasks in computer vision and audio synthesis?
19. What are some key innovations that have extended Transformer models across modalities?
20. What are the limitations of using Transformer models in real-time or resource-constrained environments?

REFERENCES

Chelliah, P. R., Dutta, P. K., Kumar, A., Gonzalez, E. D. S., Mittal, M., & Gupta, S. K. (Eds.). (2025). *Generative Artificial Intelligence in Finance: Large Language Models, Interfaces, and Industry Use Cases to Transform Accounting and Finance Processes.* John Wiley & Sons.

Creswell, A., White, T., Dumoulin, V., Arulkumaran, K., Sengupta, B., & Bharath, A. A. (2018). Generative adversarial networks: An overview. *IEEE Signal Processing Magazine, 35*(1), 53–65.

Croitoru, F. A., Hondru, V., Ionescu, R. T., & Shah, M. (2023). Diffusion models in vision: A survey. *IEEE Transactions on Pattern Analysis and Machine Intelligence, 45*(9), 10850–10869.

Cui, J., Zhu, B., Xu, Q., Tian, Z., Qi, X., Yu, B., . . . & Hong, R. (2025). Generalized Kullback-Leibler divergence loss. *arXiv preprint arXiv:2503.08038.*

Gillioz, A., Casas, J., Mugellini, E., & Abou Khaled, O. (2020, September). Overview of the transformer-based models for NLP tasks. In *2020 15th Conference on Computer Science and Information Systems (FedCSIS)* (pp. 179–183). IEEE.

Goodfellow, I., Pouget-Abadie, J., Mirza, M., Xu, B., Warde-Farley, D., Ozair, S., . . . & Bengio, Y. (2020). Generative adversarial networks. *Communications of the ACM, 63*(11), 139–144.

Kingma, D. P., & Welling, M. (2013). Auto-encoding variational bayes. *arXiv preprint arXiv:1312.6114.*

Madaan, G., Asthana, S. K., & Kaur, J. (2024). Generative AI: Applications, models, challenges, opportunities, and future directions. *Generative AI and Implications for Ethics, Security, and Data Management*, 88–121.

Mak, H. W. L., Han, R., & Yin, H. H. (2023). Application of variational autoencoder (VAE) model and image processing approaches in game design. *Sensors, 23*(7), 3457.

Mehmood, R., Bashir, R., & Giri, K. J. (2023). Deep generative models: A review. *Indian Journal of Science and Technology, 16*(7), 460–467.

Odaibo, S. (2019). Tutorial: Deriving the standard variational autoencoder (vae) loss function. *arXiv preprint arXiv:1907.08956.*

Pinheiro Cinelli, L., Araújo Marins, M., Barros da Silva, E. A., & Lima Netto, S. (2021). Variational autoencoder. In *Variational Methods for Machine Learning with Applications to Deep Networks* (pp. 111–149). Cham: Springer International Publishing.

Sadiku, M. N., Ajayi, S. A., & Sadiku, J. O. (2025). Artificial intelligence in media and entertainment. *International Journal of Scientific and Applied Research (IJSAR), eISSN: 2583–0279, 5*(5), 1–14.

Sai, S., Gaur, A., Sai, R., Chamola, V., Guizani, M., & Rodrigues, J. J. (2024). Generative AI for transformative healthcare: A comprehensive study of emerging models, applications, case studies, and limitations. *IEEE Access, 12*, 31078–31106.

5 Fairness in GenAI

When considering the concept of fairness in GenAI, one quickly encounters complex and multifaceted issues. It is not solely concerned with the elimination of bias; rather, it also encompasses the equitable representation of diverse groups, thereby ensuring fairness. The goal seems to be beyond neutral; therefore, each group gets a fair voice in what the AI creates. Therefore, it is a big deal, and these systems appear everywhere now (writing essays, sorting candidates, and even operating bots). Therefore, we must try to keep everything fair. This is not easy at all. Hidden preferences slip into the results, so the real balance is still fuzzy. Moreover, some think that we must see who actually made the rules. This bias is deep-rooted and difficult to eliminate. Even if we somehow made the training data and the answers given by AI look fair, we still have to ask what fairness really means and how we could measure it (Chiappa & Gillam, 2020). Trying to decide what is fair? You have a whole bunch of things to consider; the list gets considerably long.

Therefore, how can we sort out their fights inside each group and the battles they start with each other? It is tough, no doubt. However, if GenAI becomes fair, it could reshape society. Who knows what that might bring? Fair AI, if we ever manage to build it, might help society in health, education, and work, thereby making life easier. Maybe it opens doors for low-income teens; therefore, the gap shrinks more (could it really shift how society treats everyone?).

There could be a big jump in our money chances, so we might see more jobs popping up; therefore, our wallets get a bit heavier.

Therefore, they may allow access to many key resources quickly and easily. Still, I really doubt that we will see any of those benefits; therefore, we must first tackle the fairness problems we have already listed, must not we? GenAI is not just a shiny tech fad; it is actually kick-starting new ideas, shaping fresh products, and even handling some chores itself.

Tools such as ChatGPT, DALL·E, and Midjourney have appeared in places you would not expect, such as a newsroom, marketing firm, high-school classroom, and even a hospital. Therefore, people have started asking whether it works. Is it fair? Fairness is linked to how we perceive justice, diversity, and inclusion on the screens and apps we use. The power of these generators is massive; they produce text, images, videos, and even fake voices that look just like something a person created. Moreover, no other program does that much, and because the output can blend in, readers may be fooled without even noticing (Lundberg & Lee, 2017).

These new creations could revolutionize the opinion formation process, the cultural shifts we experience, and the very act of selection in our lives. A biased model would be a more sophisticated way of maintaining and amplifying the old divides. Some argue that it should not be used until it is better understood, while others see no reason for pause and think the benefits are worth the risks. School principals, for example, worry that AI could usher in a new era of procrastination

DOI: 10.1201/9781003713951-5

and replace the bad writing habits we have always had with bad writing habits powered by AI.

Considering fairness in GenAI, the issue intersects with big ideas from ethics, sociology, and law; therefore, the conversation gets messy. Is a system called fair when it treats all people the same, or when it tries to give each person a real chance, therefore feeling just? Could this mean that the outcomes match, the chances match, or that the voice simply remains proportional? So those old debates? Engineers and ethicists building models today cannot solve these problems (Mitchell et al., 2019).

Therefore, with the world so tightly linked, we ask: can one nation's idea of fairness become a rule for everyone? **Talking about fairness in AI that writes stuff is already confusing; actually getting it right? That is a whole other knot.** Honestly, fairness is separate from things that look alike; that is the whole deal. Whenever an AI picks something, there are winners and losers; therefore, the outcome is split. People want teachers who set the homework rules to be open about how those rules actually help us study; therefore, they should explain it plainly; otherwise, we are just guessing. **Transparency, they say, seems key for fairness; therefore, people affected can actually see why a choice was made. Clear decision: Its impact is judged fairly.** Therefore, if the people in charge are not forced to answer for bad results, then all the talk about transparency and fairness loses any real weight.

In the next section, we will check what biases show up in GenAI, and then we name the kinds we found. How can the problem be measured? We then named a few political angles that kept the discussion rolling. By the end, you should get a picture of how fairness can—and should—be part of building, using, and governing GenAI tools. This text touches on bias-reduction tricks and points to future policies.

5.1 BIAS IN TRAINING DATA DESCRIPTION

One major question regarding fairness in GenAI is the data it learns from. The training sets we provide to GenAI often copy the unfairness in our world; therefore, the output may end up being slanted or even unfair. For example, most data come from Western sources, which could make the GenAI speak in only one cultural voice. (Mehrabi et al., 2021).

Bias in GenAI appears to arise from a set of causes. The most obvious is the data. If the data are bad, the model will be bad, and the results will be biased. This point seems almost too simple. However, even with perfect data, another problem remains: humans who guide or test AI bring their own views. Therefore, the system may remain biased even in an ideal data world.

Some scholars say that data must be audited before training, while others argue that too much cleaning removes useful nuances. In conclusion, fairness is not solved by clean data alone; it also requires careful human oversight in practice.

- *Historical Bias*: Historical records often show past unfair treatment, which may push models toward making biased forecasts.
- *Unbalanced Class Distribution*: A face-generation system trained mostly on light-skinned faces appear to struggle with dark-skinned faces.

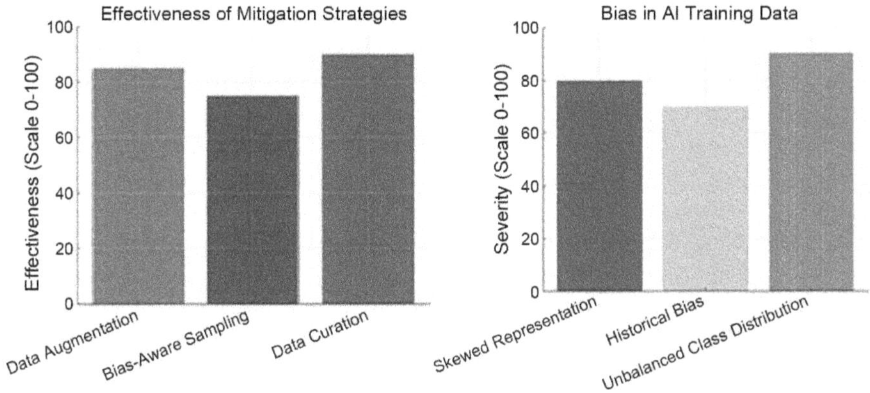

FIGURE 5.1 Bias in Distribution

This suggests that the dataset was unbalanced. Consequently, the outputs become less realistic for certain groups, revealing a clear class-distribution problem. This raises questions regarding the fairness of AI (Liang et al., 2022).

- *Mitigation Strategies*:
  - *Data Augmentation*: We ensured a better representation of the data by introducing a more diverse dataset.
  - *Bias-aware Data Sampling*: We achieve balance across various demographic groups.
- *Data Curation and Preprocessing*: We actively filtered and balanced the dataset distribution. Figure 5.1 shows bias in distribution.

A foundational concern regarding GenAI pertains to the potential biases embedded within the models. For instance, a generative image model can exhibit a marked bias in the racial and gender representation of various professions, depending on the training data. If the training data are heavily skewed toward a certain demographic, such as the profession of doctors depicted as mostly white males, the images produced may be a misleading global representation.

5.1.1 Algorithmic Discrimination

Concerns regarding bias in AI are growing. This appears to stem from the data we feed into the models and how we construct those models. The problems appear as word embedding bias, mode collapse in GANs, or models that adhere too closely to what is represented in the largest group of their training set. When this occurs, it may mean that the people whom the model is intended to serve are treated unfairly. Consider a hiring system that seems to prefer resumes constructed in a certain way, a loan program that seems to be pushing down applicants from a certain neighborhood, or a streaming service that suggests the same content to everyone (Bolukbasi et al., 2016).

One idea is to add fairness rules when training the model. Another approach is to use regularization or oversample underrepresented examples so that the system does not favor the majority. Some scientists have even designed architectures that detect bias.

More recent work, such as adversarial debiasing or reinforcement learning methods, is promising for cutting bias (Zhang et al., 2018). As AI assumes a larger role in our everyday decisions, it is vital to avoid putting our tools at risk. Even when using seemingly balanced data, a system might behave unfairly for a number of reasons that can be broadly categorized as follows: because it was programmed to be unfair; because of what it tries to optimize; or because of hidden assumptions wedded to it. Let us consider these separately.

One key contributor is the loss function optimization. Many models are set up to reduce some average error, which can be quite good for kinds of accuracy that might not show up until after the model has done its averaging. An instance that might not show up as a problem on average is unfairness impacting a minority group.

- *Bias Amplification*: Deep networks may amplify subtle biases present in the training set due to nonlinear feature interactions.
- *Example*: Text-to-image models may generate images that reflect exaggerated gender roles when asked to visualize professions like "nurse" or "CEO."

5.1.2 Bias in Human Annotations Description

Numerous AI systems depend on labeled data, which can introduce bias because the individuals providing the labels have their own views, cultures, or concealed prejudices (Snow et al., 2008).

One specific instance of this labeler bias can result in what we might term toxic bias: our labelers might be marking more tweets in African American Vernacular English as "toxic." This could cause a model to unfairly punish speech.

Another instance may involve stereotypical labeling, where our labelers marked most pictures of women in the workforce with caregiving terms. When labelers make normative judgments inconsistently or based on prejudiced views, AI systems powered by labeled data can also be expected to make biased judgments. Figure 5.2 shows comparison between bias and its effectiveness.

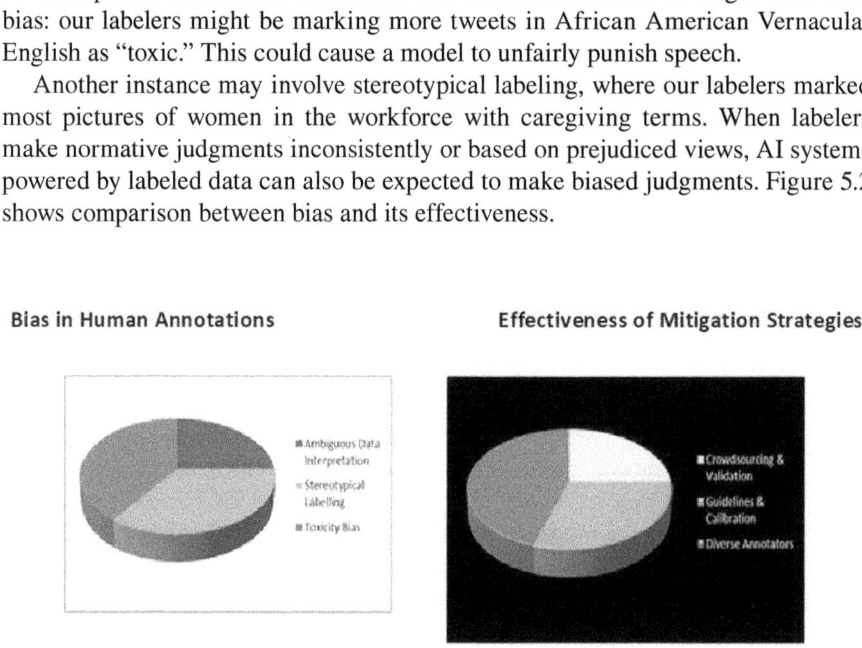

FIGURE 5.2 Bias in Human Annotation and Mitigation Strategy

Labels created by humans often reflect these hidden biases. When annotators tagged the data, they could infuse parts of their own culture, stereotypes, or unspoken preferences into the labels. For instance, is AAVE (African American Vernacular English) called "toxic" more often than Standard English? Perhaps, if so, it is biased. Pictures of women are often labeled as "caring" or "pretty" but not "professional." Is this result consistent with the number of pictures of women who have held high office?

Furthermore, inconsistencies due to cultural misunderstandings or vagueness in the guidelines can also arise. Therefore, in addition to using better labelers, what is a way to mitigate bias in labeling? One way is to talk to the labelers. Not in the way of long meetings but in the way of short setups that give the labelers a few helpful pointers on how to proceed without bias (Ouyang et al., 2022).

5.1.3 Systemic and Feedback Bias

Bias does not happen without some movement on its part to be seen and, often, to be deemed acceptable. When it comes to AI tools like ChatGPT and DALL.E, we must pay close attention to the kinds of biases that might be coded into them during their creation and the kinds of biases that might influence the feedback loops they form when they interact with users. For example, popularity bias occurs when users click on or share content that is a stereotypical image. When this happens, the model being used in the AI tool could start to see that kind of image as a good example of what to generate when asked for an image of a person.

In a way, the AI tool is starting to adopt the biases of its most prominent users. Another potential problem is the curation bias. This might occur when content filters used by the platforms that host these AI tools unintentionally mute certain voices while amplifying others.

5.1.4 Cross-Cultural and Linguistic Bias

Bias becomes evident when machines communicate across cultures. A model may struggle with underrepresented languages, leading to overly general or incorrect outputs. However, even with the number of low-resource languages and dialects considered, the picture that emerges is unsettling for diverse global communities. Live in a culture for any length of time, and you learn that all societies have a wide range of "local" or "insider" knowledge that figure prominently in their expressions, for good or ill.

Like any human, a nonhuman actor can only communicate in the ways it has been trained. No amount of "fixing" the system can account for all the potential nuances and variables that come into play when any one of us is trying to understand and make sense of our shared world. Indeed, most fixes involve feeding the system more and more data and doing more and more with different kinds of intelligences—local, regional, and otherwise—to check and recheck its output.

5.1.5 Intersectional Bias

Table 5.1 shows types of bias in GenAI

TABLE 5.1
Types of Bias in GenAI

Type of Bias	Description	Example	Possible Mitigation
Historical Bias	Bias in source data due to past inequalities	Gender pay-gap data in training	Data rebalancing
Sampling Bias	Over- or underrepresentation in dataset	Overuse of Western internet data	Stratified sampling
Annotation Bias	Subjectivity in labeling	Labeling AAVE as toxic	Diverse annotation teams
Algorithmic Bias	Model architecture preferences	Mode collapse in GANs	Fairness-aware loss functions
Feedback Bias	Bias through user-model interaction	Reinforcement of stereotypes	Continuous monitoring
Cross-cultural Bias	Bias from language or culture gaps	Poor generation in Hindi or Swahili	Regional fine-tuning
Intersectional Bias	Overlapping marginalization	Misrepresentation of Black LGBTQ+ users	Intersectional audits

5.2 APPROACHES TO PROMOTE FAIRNESS

One way to assess whether AI treats all humans fairly is to examine fairness metrics. This is an important aspect that should be considered in the design of AI. Furthermore, it can and should be studied using datasets that are not only diverse but also representative of the user population. This is because if the population is underrepresented in the dataset, the way AI divides or conquers the dataset will not produce a fair result (Hutchinson et al., 2020).

5.2.1 Algorithmic Debiasing Techniques

Bias in models can be reduced through adversarial debiasing and reweighting. Achieving transparency and maintaining thorough documentation can aid this effort. These practices allow us to record not only our data sources but also the model design and the rationale for our decisions at numerous steps along the way, making all of it reviewable by outsiders. Along with these practices, audit processes should be implemented, the need for which I have ample reason to believe is long overdue. Fairness problems and model bias, more generally, should have "fix over time" mechanisms built into them.

5.2.2 Promoting Equitable Access

Can equitable AI create an accessible and efficient healthcare system? This question is critical because the answers may also bring schools and employment into the efficient technology fold.

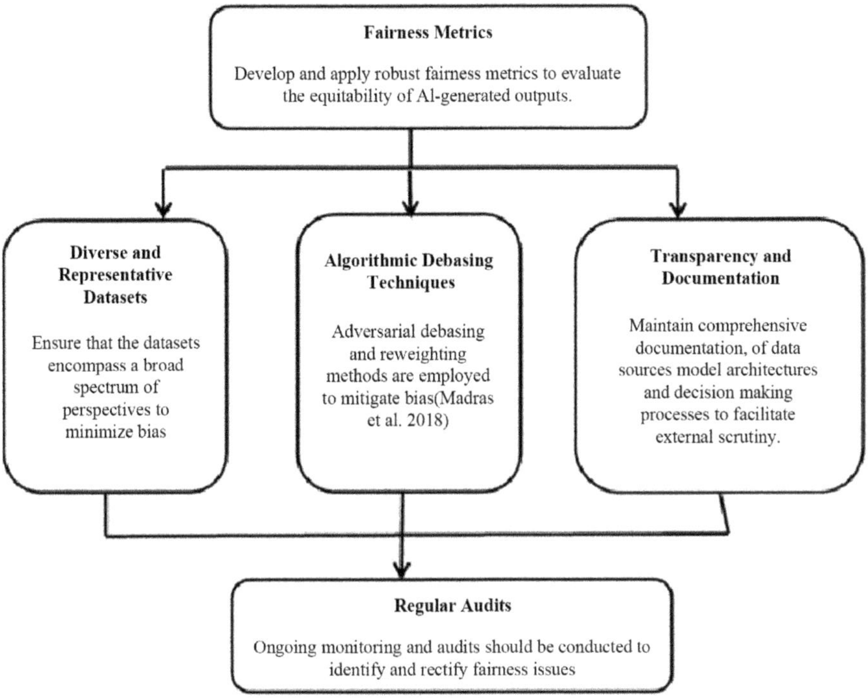

FIGURE 5.3 Approaches to Promote Fairness

Thus, using fair AI in these sectors may push us toward an ethical technology future. Moreover, when companies use AI that is fair, they may also comply with the many legal and moral hazy areas that technology has fallen foul of in the past. Figure 5.3 shows approaches to promote fairness.

5.3 FAIRNESS MEASURING TECHNIQUES

Fairness in GenAI feels like a colossal puzzle, necessitating careful consideration of and solutions to several persistent issues. For years, we have tried various methods for measuring fairness, with some researchers now claiming AI is "fair" when it achieves something called "demographic parity." In actuality, demographic parity—a fancy term for equal representation—happens to be a method for measuring fairness that does not account for the kind of errors AI systems can make (and currently, they are making a lot). Consider a system that makes the same type of erroneous decisions across the board. Should we celebrate this as fair; of course not. We should want a system that makes more good decisions than bad, with more good decisions made consistently across the groups it is being asked to serve. Counting human and digital images, as well as the kinds of decisions AI makes, is difficult to do with any level of artistic merit. Should we pay homage to our idols and ask everyone to keep it a secret? Or do we try to imagine the Western United States in a way that feels fresh to us every moment of the day? Ultimately, ensuring that GenAI is fair, that is, not biased against certain groups, is the responsibility of the humans who create and deploy it.

Current tools, such as Shapley Additive Explanations (SHAP), Local Interpretable Model-Agnostic Explanations (LIME), and IBM AI Fairness 360, provide some assistance. They do not "make" models fair. They are not a magic fix. They require a clear-headed understanding of what bias is, whence it comes (e.g., in the training data), and what the consequences are for the lives of those whose paths will cross the models.

These tools can help with this understanding. They can also aid in getting us to some place that feels more fair than not. However, fairness is not a destination that we arrive at and then stop. It requires constant monitoring, various types of checks, and ongoing human commitment. Figure 5.4 shows bias in GenAI.

These instruments illuminate the decision-making processes of AI, rendering transparency that was previously absent. For the first time, researchers can now see not only the "what" of AI outcomes but also the "why" behind them.

This has clear implications for singing the gospel of AI ethics and bias mitigation. Undiagnosed and unaddressed biases can only increase. While understanding motives is beneficial, it is even more essential to understand the source of biases in GenAI. Like all biases, these biases do not spring from one source or another; they are much more complex. Because they are trained to mimic the statistical "truths" of the training data, the models will naturally tend toward "buzzy" outcomes when the training data are flawed, insufficient, or biased. A summary of fairness techniques and approaches is shown in Table 5.2.

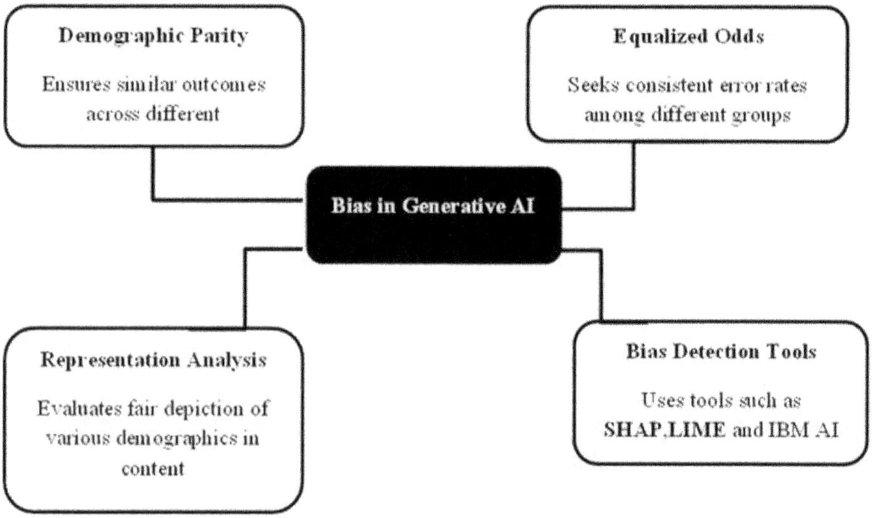

FIGURE 5.4 Biases in GenAI

TABLE 5.2
Approaches, Techniques, and Implications of Fairness in GenAI

Category	Approach/Technique	Purpose/Benefit
Fairness Metrics	Demographic Parity, Equalized Odds, Counterfactual Fairness	Assess equity in AI-generated outcomes across demographic groups
Dataset Diversity	Use diverse and representative datasets Hutchinson et al. (2020)	Minimize bias in training data
Algorithmic Debiasing	Adversarial debiasing, reweighting methods Madras et al. (2018)	Mitigate bias during model training and prediction phases
Transparency and Documentation	Maintain detailed logs of model decisions, data sources, architecture	Allow for external review and accountability
Bias Detection Tools	SHAP, LIME, IBM AI Fairness 360	Identify and explain sources of bias in models
Representation Analysis	Evaluate demographic representation in generated content	Prevent underrepresentation or misrepresentation
Ongoing Audits	Continuous monitoring and fairness audits Raji et al. (2020)	Detect and correct bias over time
Ethical Policy Integration	Apply fairness policies and governance standards	Ensure compliance and responsible innovation

5.4 CASE STUDIES OF BIAS IN GENAI

5.4.1 Text Generation Bias: GPT in Hiring Simulations

Gender bias in GPT-3 seems to come from the internet training set, which embodies all kinds of troublesome biases and stereotypes—as well as the desirable characteristics—that people have about one another. Models like Bender et al. (2021) show quite clearly that if people assign inherently positive or negative attributes to one gender or the other, the model is quite likely to replicate that kind of thinking.

It is not a deliberate code choice; the model simply copies the patterns it sees. Because blogs, ads, news, and even movie scripts show the same biases, AI repeats them. Ask it about CEOs and the output usually sounds male; ask about nurses and it leans toward female. This raises a big question for developers: can we achieve fairness if the data are not filtered? Some advocates contend that training samples should be more balanced and that bias-reduction techniques should be used. Others maintain that our world is always somewhat prejudiced and that real-world data can never be truly neutral. They assert that even the most meticulous data cleansing can never eliminate all the hidden biases built into our assumptions about the value and proper use of gender across different cultures.

5.4.2 Image Generation Bias: DALL·E and Occupational Stereotypes

Birhane et al. (2022) showed that large image generators, such as DALL·E 2, often replicate the biases of race and gender found in their training data. When the prompt is as simple as "CEO" or "engineer," these systems default to producing images of white men. This is not the result of some sneakily added programmer bias; it is spreading the way all internet biases spread—from huge datasets filled with social inequality and stereotypes. These visual shortcuts can reinforce and perpetuate harmful ideas about who should and should not be considered for certain careers.

Attempts to fix this situation by adjusting the prompts, as OpenAI has done, are at best temporary Band-Aids that do not address the fundamental problem with the model itself. They do not change the core of the model, which still reacts to biased or vague queries. Therefore, unless users specifically tell the system to think differently, it may continue to produce skewed pictures.

The research suggests that technical patches alone are insufficient. We likely need larger ethical guidelines and possibly redesigning training data. Only then can visual AI become fair for everyone. This would help create images that everyone trusts.

5.4.3 Chatbot Conversations: Reinforcement of Racial Stereotypes

Chatbots that learn from sites such as Reddit or Twitter may end up carrying the same biases that appear there. These platforms have many voices, but the loud ones often drown out the quiet ones.

Therefore, it is quite possible that a model trained on raw posts will pick up hate or unfair remarks without even trying. Solaiman and his team (2019) observed that these systems can soak up toxic language that is common online. This is concerning because these agents are supposed to talk politely with anyone. Could they unknowingly perpetuate harmful stereotypes? Ganguli et al. (2022) also mentioned that even when developers add human feedback, the bias problem may stay if the feedback itself is not checked well. If most reviewers share the same background, the system will likely repeat that view and ignore the others.

This indicates that more transparent methods are required to display the construction of models. Gathering data from a more extensive and varied set of individuals could aid our efforts. Certainly, running "fairness checks" could also be part of a solution, but finding more open and frequent means to signal and keep our models honest is perhaps the larger lift we need to accomplish. This is necessary if we hope to prevent conversational AI from continuing to reproduce (and perhaps amplify) unfair human behaviors and biases.

5.4.4 Educational Tools: Gendered Language in Writing Assistants

In 2018, Google seemed to stop providing suggestions with gendered pronouns in Gmail's Smart Compose after the tool was found to exhibit a steady bias in the guesses it made. Specifically, the tool often defaulted to a male pronoun when it did not know the gender of the person the writer was referring to. Of course, it did this even when a large number of the people being written to were female. This problem is not limited to autocomplete, as you have probably guessed (Caliskan et al., 2017).

This is something that is being implicated in large pretrained language models in general. For instance, in a paper by Aylin Caliskan, Almond Bryson, and Arvind

Narayanan, these researchers reported that a large pretrained model was much more likely to associate "brilliant" and "logical" with male names and "nurturing" and "supportive" with female names.

5.4.5 Healthcare Applications: Underrepresentation in Medical Generative Models

Obermeyer et al. (2019) point out that medical AI can end up copying existing health gaps when the data used miss big racial groups. Their study found that a popular health algorithm tended to lowball the needs of Black patients compared to White patients. This occurred because the model used health spending as a sign of condition but ignored the fact that minority groups often receive less care. This may mean that the algorithm simply follows the bias in the data.

Later, Seyyed-Kalantari et al. (2021) showed that image-based diagnostic tools have larger error rates for minority people, likely because these faces were few in the training set. The results appear to be less reliable for these communities, which could cause incorrect diagnoses or treatment. Banerjee et al. (2021) examined medical GPTs that write patient summaries. They noticed that conditions common in underrepresented groups were described vaguely or were missing.

This type of omission probably hinders communication and could lead to poor treatment plans. Overall, these studies suggest that we need more diverse, representative data and tests that look for bias. Without this, AI may continue to treat people unequally, which is simply not acceptable. Future studies could help improve the fairness for all patients. Table 5.3 summarizes case studies on bias in AI systems.

5.5 METRICS TO EVALUATE FAIRNESS IN GENAI

Measuring fairness in GenAI may seem important, although we have not yet determined the best way to do so. Formal metrics can provide numbers that indicate whether a system treats people or groups equitably. These numbers help developers shape models so that the results do not maintain or raise social gaps, especially in the health field. Classic fairness scores from classification do not fit; they require tweaks for each type of output: text, picture, or sound. This section lists the main fairness measures and their significance for the generation tasks. Policymakers should also consider these factors.

5.5.1 Demographic Parity

Demographic parity, or statistical parity, means that a model's output should not depend on traits such as gender, race, or age. In GenAI, this could mean that the system creates pictures or text without favoring any group. For example, a request for a "professional" portrait that mainly shows White men would break this rule. Some researchers believe that this is difficult to achieve because the data often reflect bias in practice.

However, the concept is to maintain the probability of yielding a positive result equally among all demographic groups. Hence, evaluating the models may

TABLE 5.3
Case Studies on Bias in AI Systems

Study	Domain	Key Findings	Implications
Bender et al. (2021)	Language Models (GPT-3)	Leadership prompts yielded male-oriented responses; caregiving prompts aligned with feminine characteristics.	Reveals how gender-role stereotypes are replicated due to biased training data from the internet.
Birhane et al. (2022) Ramesh et al. (2022)	Image Generation (DALL·E 2)	Neutral prompts like "CEO" or "engineer" resulted in White, male images; OpenAI applied prompt augmentation to reduce bias.	Visual models reinforce racial and gender stereotypes; prompt-level fixes are limited.
Solaiman et al. (2019) Ganguli et al. (2022)	Conversational AI (e.g., Reddit-based models)	Chatbots inherit biases from dominant groups in training communities; RLHF signals may further reinforce bias.	Highlights the need for fairness auditing and diverse feedback in RL processes.
McMillan (2018) Caliskan et al. (2017)	Autocomplete and Word Association	Gmail defaulted to male pronouns in ambiguous contexts; language models linked "brilliant" with male names.	Gender bias in predictive text tools; led Google to disable gendered pronoun suggestions.
Obermeyer et al. (2019)	Healthcare Algorithms	Health risk algorithm underestimated needs of Black patients due to cost-based proxy for health.	Reinforces racial disparities; underscores flawed assumptions in training proxies.
Seyyed-Kalantari et al. (2021)	Diagnostic Image Models	Higher error rates observed for minority patients due to underrepresentation in training data.	Increases risk of misdiagnosis in marginalized communities; calls for more diverse data.
Banerjee et al. (2021)	Medical GPT Systems	Clinical summaries lacked specificity for conditions common in underrepresented groups.	May impair treatment quality; stresses need for inclusive training in clinical NLP models.

necessitate ongoing review and recalibration to ensure that they perform equitably across all groups (Hardt et al., 2016).

$$P(\hat{Y}=1|A=0) = P(\hat{Y}=1|A=1)$$

Demographic parity attempts to force different groups to achieve similar outcomes, but it may not consider the actual, sometimes significant, differences between groups. This goal could be achieved even if it comes at the expense of the accuracy of the model—sacrificing predictive performance for a fairer appearance. Indeed, a model could score high on demographic parity even if it were no better than chance in making accurate predictions. At present, there is no way to know which models of meritocracy achieve this goal by means of natural variation and which do so by means of bias.

5.5.2 Equalized Odds

Equalized odds are a fairness concept that states that a model should maintain the same true-positive and false-positive rates for different groups. In the world of GenAI, this could mean that the system works equally well and makes the same kind of mistakes for everybody. Consider a GenAI that assists doctors in writing diagnoses. It should not be more right for one ethnic group than for another. The same goes for caption generators or essay graders; the output should not give credit or hurt anyone because of their race or gender.

This measure is important in domains such as health, school, or money, where a biased slip can cause real harm. However, incorporating equalized odds into a generative tool is not straightforward. The output often has no single right answer, so counting errors can be tricky. Therefore, careful human checks and benchmarks are also required. Some argue that excessive testing can slow progress, but fairness probably requires that effort.

5.5.3 Counterfactual Fairness

Counterfactual fairness is a standard for evaluating fairness that examines whether an AI model's results would remain consistent if a person's sensitive characteristic—such as gender, race, or age—were changed, while all other individual attributes remained constant (Kusner et al., 2017). This criterion is met when the predicted result for a person with the actual attribute value ($A = a$) matches the predicted result if the attribute is different ($A = a'$), expressed as:

$$\text{Fairness} \Leftrightarrow YA{\leftarrow}a = YA{\leftarrow}a'.$$

In the realm of GenAI, this concept aids in the detection of subtle and indirect biases. For instance, if a resume creation tool uses more assertive or leadership-focused language for a male input than for a female one with the same qualifications, the system does not pass the counterfactual fairness test.

Likewise, in applications such as personalized education or storytelling, altering a subject's race or gender should not change the tone, content, or quality of the generated narrative unless such changes are contextually warranted. Implementing counterfactual fairness involves understanding and modeling the causal relationships between variables, which is technically challenging. Nonetheless, it offers a valuable perspective for identifying and addressing biases that might not be apparent through statistical analysis alone, particularly in personalized or content-rich GenAI systems.

5.5.4 Individual Fairness

Individual fairness attempts to treat similar people with similar outcomes (Dwork et al., 2012). In generative work, this may mean that two similar prompts receive equally clear or upbeat replies, regardless of who writes them. Judging similarity requires a clear method, which is a necessary condition when the output itself is personal, such as a tutoring note or a bank report written by an AI. However, some critics argue that it can be very difficult, if not impossible, to agree on a function for judging the similarity between inputs and outputs for the learner or the client.

5.5.5 Representation Parity and Coverage Metrics

Generative models that invent text, images, or other novel media cannot be fully evaluated based on accuracy alone. Instead, the focus should be on how well they represent the astounding variety of humans across demographic groups. The metrics used for this type of focus are called representation metrics.

They are important in applications such as image generation, storytelling bots, and virtual assistants. What we see can shape our thoughts on inclusion. One common metric is the representation parity. It checks whether marginalized groups (such as racial or ethnic minorities) are represented in the output as often as we would hope. If a picture generator, for instance, is set to draw "leaders," and it gives us an output with nine White men to one output of a woman, parity seems poor. This is likely to send a signal that women are underrepresented and, by and large, are missing from the conversation about leadership. Simply having diverse representation and assigning various roles is insufficient to ensure fairness in portrayals. The problem arises, obviously, when one has been sidestepped to make room in the public consciousness for the other.

5.5.6 Embedding and Latent Space Fairness

Fairness work has gone inside the black box of GenAI to examine the bias in the model's hidden layers. These layers are where the model stores its collective "knowledge" and are critically important for large foundation models, such as those that create text and images. Foundation models learn to share their internal representations across a wide variety of data types. When probing these models for fairness, we must first look for bias in the representations they learn (Bolukbasi et al., 2016). Approaches such as the Word Embedding Association Test (WEAT) allow us to do this.

The WEAT checks how tightly or loosely certain "protected" human traits (like gender) are bound up with even "neutral" words. We also used WEAT to determine which words are more or less likely to be selected as examples when performing a task. Fairness work also examines the shapes of embeddings for different social groups. For this, we used manifold alignment techniques and sometimes just good old eyeballing. If we find that the embeddings of two different groups are shaped very differently, it is a sign of unfairness.

5.5.7 Tools for Fairness Measurement in GenAI

Amid growing concerns over bias in AI, researchers have begun to create toolkits to check the fairness of generative models. One of the most well known is AIF360, which was created by IBM. Similar to many of its kind, AIF360 provides a number of scores and "fixes" for bias (Bellamy et al., 2019). It was first used on tabular data but has since been modified to work with more data types, including text and images, resulting from generative models. Another such toolkit is Fairlearn, which was developed by Microsoft. Fairlearn attempts to show the trade-offs between a model's accuracy and fairness. Both AIF360 and Fairlearn can be plugged into ML pipelines, although they are not the only tools available. SHAP and LIME are not fairness tools, but they provide local explanations that some researchers use to determine how inputs to a model might cause it to produce unfair or biased outputs.

5.6 LEGAL, ETHICAL, AND POLICY DIMENSIONS OF FAIRNESS IN GENAI

Ensuring that GenAI is fair is not just a numbers game; it is also about ethics and law. As these systems find their way into schools, hiring and policing, and even news, the chance for harm seems to grow. If they make mistakes, we could be in for a world of hurt when it comes to bias (the kind that already has some governors using AI to assess our schools in very questionable ways), reputational damage (if an AI system wrongly accuses someone of something, it could be a bit like the time someone accused a high school baseball team of using steroids), and even human trust (these systems are talking to us, with our voices, after all).

5.6.1 Legal Frameworks and Regulatory Trends

Governments are concerned about bias in AI and have started to take action; rule-making, in the form of legal frameworks, policy guidelines, and oversight committees, could, if done well, ensure AI's fairness in practice.

5.6.1.1 The EU AI Act

The EU's AI Act may categorize generative tools used for hiring and in educational settings as high risk. This may necessitate stricter regulations. As with much of the EU legislation, the Draft Act prioritizes transparency. The AI Act will also require that the systems it covers have "data integrity," which means that the inputs to AI systems (including generative tools) must be of high quality. "Bias evaluation" is a phrase that has come up a lot in the past several months, and for good reason (European Commission, 2021).

5.6.1.2 US Algorithmic Accountability Act

This legislation, introduced in 2022, mandates the evaluation of the impact of AI systems on human rights, health, and finances. This emphasizes the necessity of providing reasonable assurances that the systems are fair, that they operate with the kind of transparency that is essential for accountability, and that protective measures are in place to avoid causing any kind of harm to the users. The legislation encourages the responsible design and use of AI systems in sensitive areas (Schwartz & Stark, 2022).

5.6.1.3 NYC Local Law 144 (2023)

In New York City, a new law requires annual bias assessments of automated hiring systems. Local Law 144 is the first enforceable fairness regulation for AI in the United States. Local law compels employers who use AI in recruitment to address concerns over discrimination and heighten transparency (Raji et al., 2022).

5.6.1.4 Global Data Protection Laws

International regulations, such as the GDPR in the EU, Personal Information Protection and Electronic Documents Act (PIPEDA) in Canada, and Digital Personal Data Protection (DPDP) in India, oversee the use of AI when it involves personal data. The GDPR's right to explanation may imply that AI choices must be clear. This seems to help fairness by shielding users from automated content and privacy.

5.6.2 Core Ethical Principles in GenAI Fairness

Ethical principles form the foundation of responsible AI. Key principles relevant to fairness include the following:

5.6.2.1 Justice and Nondiscrimination

People should not be treated unfairly because of their race, gender, or age. Involving data that some would say is neutral, but that is very often not, we should insist upon fairness rules and checks—our GenAIs should not be harmful to humans. It is us, the creators of the AIs, who are responsible for the harmful aspects of the AIs. Output checks for harmful content should be a basic part of the pre-release process. This is aligned with the ethics of Floridi et al. (2018).

5.6.2.2 Autonomy and Consent

Individuals can maintain their freedom when they guide AI in making decisions on their behalf, especially for matters such as selecting narratives or products for themselves. Consent should be the ethical underpinning of this. When "AI" is an "I" making a copy of a real person, as in the case of a deepfake, the entity in the video should and must be backstopped with a decision on whether or not he or she gives permission for that scenario to unfold. Life, as it happens in a "post-truth" world for social media or otherwise, should have these nonsensical situations governed at least somewhat sensibly by balance.

5.6.2.3 Beneficence and Non-Maleficence

Beneficence seems to push AI to lift user well-being, while non-maleficence just wants it not to hurt. For a GenAI, this might mean ensuring that answers remain helpful, fair, and responsible, especially in fields such as medical care, mental support, or schools. We probably need close checks on such models so that false information does not spread, feedback is not stifled, and vulnerable people stay safe to protect future generations everywhere.

5.6.2.4 Transparency and Accountability

Keeping a GenAI system open may allow people to see how it works and why it provides certain answers. This could mean having simple paperwork, such as model cards or data sheets, plus obvious warnings in any text it creates. Likewise, accountability may require a way to hear complaints, fix any damage, and ensure that makers and runners answer ethical issues. Therefore, both openness and responsibility seem to be essential in practice today.

5.6.3 Risks of Non-Compliance and Ethical Negligence

The unfairness of GenAI can lead to societal and legal problems. One of the main concerns is the biased output. A job advertisement written by AI could, without any indication of intent, discourage women from applying for the position. The possible effects of such bias on company reputation are worrisome too; the tool could get called "skewed" for good reason, and a firm that leans on it might take some hits to its trustworthiness and see its brand going wobbly.

It does not take much imagination to consider the kinds of legal consequences that could ensue for a company that ignores necessary bias checks and is then accused of writing discriminatory ads. Of course, this is a concern for the potential downsides of GenAI. On the upside, some argue that it could help address issues of bias if used correctly (Binns, 2018).

5.6.4 Toward a Governance Model for Fair GenAI

AI is complex and powerful, and like anything powerful, it can cause harm. Some expect it to cause harm, and they may have good reason to worry. A clear governance plan for ethical and legal rules must be established and followed to ensure that AI can be trusted to be fair. One way to do this is through what you might call "iambic pentameter," which, strangely enough, relates to one of the not-so-secret ways researchers and practitioners keep their AIs in line. "I" stands for involvement (Veale et al., 2018). For AI to be unfair to a group of people, it generally means that the people involved in creating or testing it either did not know it was unfair or, more nefariously, did know and went ahead and made it anyway.

To prevent both types of situations from occurring as much as possible, the design of the AI should be participatory. This means making the communities that will use AI part of the crucial decisions related to creating, testing, and governing it. Together, these practices build a solid governance framework that promotes responsible and equitable use of GenAI.

EXERCISES

Sources of Bias: Data, Algorithms, Annotations

1. What are the typical sources of bias in the GenAI models?
2. How can training data lead to unfair or skewed outputs in generative systems?
3. In what ways do annotation practices and human labeling introduce biases?
4. How do algorithmic design choices amplify or mitigate the inherent biases in generative models?
5. What are the implications of biased outputs for high-stakes GenAI applications?

Algorithmic Discrimination and Mitigation

6. What constitutes algorithmic discrimination, and in what ways does it appear in GenAI?
7. How might generative models inadvertently perpetuate societal stereotypes?
8. What are some effective strategies for addressing algorithmic discrimination within generative systems?
9. What obstacles arise when implementing mitigation techniques that do not compromise model performance?
10. How does the concept of fairness vary across different cultural, social, and legal contexts?

Techniques: Reweighting, Debiasing, Fair Datasets

11. How does reweighting training data contribute to reducing the bias in generative models?
12. What role do debiasing algorithms play in enhancing fairness during and after model training?
13. What defines a "fair dataset," and how can it be constructed for generative learning?
14. How do pre-, in-, and post-processing methods differ in their approaches to bias mitigation?
15. Are fairness techniques universally applicable, or must they be tailored for specific domains?

Measuring Fairness: Demographic Parity, Equalized Odds

16. What are the key metrics for evaluating fairness in GenAI models?
17. How is demographic parity determined, and what are its limitations in generative tasks?

18. What are equalized odds, and how do they compare to other fairness measures?
19. How do fairness metrics differ when applied to classification and generation tasks?
20. Are there trade-offs among fairness, accuracy, and model complexity in GenAI systems?

REFERENCES

Banerjee, I., Chen, M. C., Lungren, M. P., & Rubin, D. L. (2021). Radiology report generation using generative adversarial networks and reinforcement learning. Medical Image Analysis, 70, 101996.

Bellamy, R. K., Dey, K., Hind, M., et al. (2019). AI Fairness 360: An extensible toolkit for detecting, understanding, and mitigating unwanted algorithmic bias. IBM Journal of Research and Development, 63(4/5), 4–1.

Bender, E. M., Gebru, T., McMillan-Major, A., & Shmitchell, S. (2021). On the dangers of stochastic parrots: Can language models be too big? In Proceedings of the 2021 ACM Conference on Fairness, Accountability, and Transparency (pp. 610–623).

Binns, R. (2018). Fairness in machine learning: Lessons from political philosophy. In Proceedings of the 2018 Conference on Fairness, Accountability and Transparency (pp. 149–159).

Bird, S., Hutchinson, B., Kenthapadi, K., Kuppuswamy, H., Mitchell, M., & Liao, Q. V. (2020). Fairlearn: A toolkit for assessing and improving fairness in AI systems. arXiv preprint arXiv:2010.04053.

Bolukbasi, T., Chang, K. W., Zou, J. Y., Saligrama, V., & Kalai, A. T. (2016). Man is to computer programmer as woman is to homemaker? Debiasing word embeddings. arXiv preprint arXiv:1607.06520.

Caliskan, A., Bryson, J. J., & Narayanan, A. (2017). Semantics derived automatically from language corpora contain human-like biases. Science, 356(6334), 183–186.

Chiappa, S., & Gillam, T. (2020). Path-specific counterfactual fairness. In Proceedings of the AAAI Conference on Artificial Intelligence, 34(4), 3603–3610.

Dwork, C., Hardt, M., Pitassi, T., Reingold, O., & Zemel, R. (2012). Fairness through awareness. In Proceedings of the 3rd Innovations in Theoretical Computer Science Conference (pp. 214–226).

European Commission. (2021). Proposal for a Regulation Laying Down Harmonised Rules on Artificial Intelligence (Artificial Intelligence Act). https://eur-lex.europa.eu/legal-content/EN/TXT/?uri=CELEX%3A52021PC0206

Floridi, L., Cowls, J., Beltrametti, M., et al. (2018). AI4People: An ethical framework for a good AI society: Opportunities, risks, principles, and recommendations. Minds and Machines, 28(4), 689–707.

Hardt, M., Price, E., & Srebro, N. (2016). Equality of opportunity in supervised learning. In Advances in Neural Information Processing Systems (pp. 3315–3323).

Kusner, M. J., Loftus, J., Russell, C., & Silva, R. (2017). Counterfactual fairness. In Advances in Neural Information Processing Systems (pp. 4066–4076).

Lundberg, S. M., & Lee, S.-I. (2017). A unified approach to interpreting model predictions. In Advances in Neural Information Processing Systems (pp. 4765–4774).

McMillan, R. (2018). Google bars gender-based pronouns from new AI tool. The Wall Street Journal. https://www.wsj.com/articles/google-bars-gender-based-pronouns-from-new-ai-tool-1543337580

Mehrabi, N., Morstatter, F., Saxena, N., Lerman, K., & Galstyan, A. (2021). A survey on bias and fairness in machine learning. ACM Computing Surveys, 54(6), 1–35.

Mitchell, M., Wu, S., Zaldivar, A., et al. (2019). Model cards for model reporting. In Proceedings of the Conference on Fairness, Accountability, and Transparency (pp. 220–229).

Obermeyer, Z., Powers, B., Vogeli, C., & Mullainathan, S. (2019). Dissecting racial bias in an algorithm used to manage the health of populations. Science, 366(6464), 447–453.

Raji, I. D., Smart, A., White, R. N., Mitchell, M., & Gebru, T. (2020). Closing the AI accountability gap: Defining an end-to-end framework for internal algorithmic auditing. In Proceedings of the 2020 Conference on Fairness, Accountability, and Transparency (FAT*) (pp. 33–44). Association for Computing Machinery (ACM). https://doi.org/10.1145/3351095.3372873

Seyyed-Kalantari, L., Zhang, H., McDermott, M., Chen, I. Y., & Ghassemi, M. (2021). Underdiagnosis bias of artificial intelligence algorithms in chest radiographs. Nature Medicine, 27(12), 2176–2182.

Snow, R., O'Connor, B., Jurafsky, D., & Ng, A. Y. (2008). Cheap and fast—but is it good? Evaluating non-expert annotations for natural language tasks. In Proceedings of the Conference on Empirical Methods in Natural Language Processing (EMNLP) (pp. 254–263). Association for Computational Linguistics (ACL).

Solaiman, I., Brundage, M., Clark, J., et al. (2019). Release strategies and the social impacts of language models. arXiv preprint arXiv:1908.09203.

Veale, M., Van Kleek, M., & Binns, R. (2018). Fairness and accountability design needs for algorithmic support in high-stakes public sector decision-making. In Proceedings of the 2018 CHI Conference on Human Factors in Computing Systems (pp. 1–14).

Zhang, B. H., Lemoine, B., & Mitchell, M. (2018). Mitigating unwanted biases with adversarial learning. In Proceedings of the 2018 AAAI/ACM Conference on AI, Ethics, and Society (pp. 335–340).

2022

Birhane, A., Prabhu, V., & Kahembwe, E. (2022). Multimodal datasets: Misogyny, pornography, and malignant stereotypes. In Proceedings of the 2022 ACM Conference on Fairness, Accountability, and Transparency.

Ganguli, D., Askell, A., Bai, Y., et al. (2022). Red teaming language models to reduce harms: Methods, results, and lessons learned. arXiv preprint arXiv:2209.07858.

Liang, P., Bommasani, R., et al. (2022). Holistic evaluation of language models. Center for Research on Foundation Models (CRFM), Stanford University. https://crfm.stanford.edu/helm

Ouyang, L., Wu, J., Jiang, X., et al. (2022). Training language models to follow instructions with human feedback. arXiv preprint arXiv:2203.02155.

Raji, I. D., Scheuerman, M. K., & Buolamwini, J. (2022). The fallacy of AI function creep: Contextualizing the regulatory landscape of New York City's hiring algorithm law. In FAccT '22: Proceedings of the 2022 ACM Conference on Fairness, Accountability, and Transparency.

Ramesh, A., Dhariwal, P., Nichol, A., Chu, C., & Chen, M. (2022). Hierarchical text-conditional image generation with CLIP latents. arXiv preprint arXiv:2204.06125.

Schwartz, A., & Stark, L. (2022). Algorithmic impact assessments under the Algorithmic Accountability Act of 2022. Data & Society Policy Briefs.

6 Opening the Black Box
Explainability in GenAI

Explainability in GenAI is becoming increasingly important as systems transition from recognizing patterns in data to creating new content on their own, such as images, text, audio, or even molecular structures. In contrast, traditional ML decisions can usually be traced through the interpretable features they use. The most popular GenAI models today, such as GANs, VAEs, and LLMs, are black boxes. Even ethical issues such as accountability, trust, or fairness are not clear-cut, especially when GenAI is applied in high-stakes situations such as healthcare, the criminal justice system, or the media. Improving our understanding of GenAI outputs in a way that benefits society is the task set before us. Inherently interpretable GenAI architectures are one way to achieve this goal. The second is to establish standardized and meaningful ways to evaluate the systems already in place (Wachter et al., 2017).

Without major breakthroughs in this domain, GenAI may not see GenAI used in vital real-world settings. People are concerned about how well GenAI can decide or cannot decide when given a task. Some believe that the downright mysterious ways in which GenAI makes decisions render it too dangerous to be trusted (Gilpin et al., 2018).

6.1 IMPORTANCE OF EXPLAINABILITY IN GENAI

Transparency is crucial for the proper functioning of GenAI systems. It goes far in ensuring that the processes are correct. It is easy to see how this works in GenAI systems, such as LLMs, which generate simple outputs, such as text or code.

However, it is difficult to understand how the processes work when GenAI systems generate more intricate outputs, such as images or music. Misleading of people is dangerous, and AI is so powerful that if it goes wrong, it can go catastrophically wrong. Therefore, the case for transparency in GenAI systems is urgent and must be made very strongly (Goodman & Flaxman, 2017).

GenAI systems are being created and implemented continuously; however, we do not have a reliable ethical framework for overseeing them, in part because we cannot see inside them and in part because the kinds of content they create raise stunningly complicated societal questions. GenAI models can generate any number of intricate outputs, and the society to which those outputs are delivered has the right to understand the whys and hows of the model's content generation. Otherwise, we will have a situation where society is asked to swallow a whole lot of AI-generated content without even a pretense of comprehensibility.

The concept of explainability varies across the stakeholder spectrum. For developers and ML engineers, explainability is related to understanding the internal

DOI: 10.1201/9781003713951-6

workings of a model. This means knowing which bits to pull and push to get the model to generate a desired output, as well as knowing which bits not to mess with if the model is to continue producing trustworthy results (Guidotti et al., 2018).

For regulators and policymakers, auditability is practically synonymous with explainability and concerns the capacity to evaluate systems for compliance, fairness, and everything else that those systems should do (Weller, 2019).

Explainability is critical in GenAI because generative models are intrinsically transparent, and their outputs are potentially high-stakes. For GenAI to serve society in a beneficial way, it is essential to understand its decision-making process, that is, how it goes from the inputs given to the outputs received. Currently, we are unable to do this. Worse, GenAI outputs might be used in a way that is misleading, socially harmful, or otherwise ethically questionable. If we do not understand how GenAI operates, how can we ensure that it operates in a socially beneficial way?

Ultimately, explainability is not merely a technical metric. It is also a core part of the principle that GenAI should be incorporated into real-world applications ethically and successfully.

6.1.1 Trust and Transparency

GenAI systems are being deployed to perform many essential tasks in key areas of life, including healthcare, law, and finance. However, they are not given a free pass. When it comes to the work they do in these domains, it is absolutely crucial for the humans they interact with to comprehend why they produce the outputs they do and why they function the way they say they do. This is where the aught of explainability comes into play. Trust follows the comprehensibility.

There is no way to develop trust in a model's output if one cannot access its internal workings and understand how it produces that output. When stakeholders understand the rationale behind a model's suggestion—be it a medical diagnosis, legal clause, or financial report—they are far less likely to worry about that suggestion and far more likely to trust that it knows what it is doing (Radford et al., 2016).

This is a positive development. Evaluating the model's output with some healthy skepticism can help catch cases where the model is providing an output that may just be a reflection of some biased training data (Samek et al., 2017). Occasional errors in advice or assistance should be considered warnings. Any institution using a GenAI system precedes its operational decision-making with a human-level review as part of an "AI accountability" safeguard (Davenport & Ronanki, 2018). The Health and Human Services (HHS) is not the first agency to use a review process.

6.1.2 Ethical Considerations

GenAI models, especially those utilizing extensive language frameworks, are susceptible to producing content that is offensive, biased, or factually inaccurate.

Models are frequently trained using problematic data. Such data could easily be the vast collections of materials we get from the internet. These collections may include all kinds of (possibly domain-relevant) stereotypes, discriminatory language, or just plain wrong stuff that (as far as we know) has not been verified by any credible sources and does not pass any kind of reasonable reality check (occasional lies, fishing expeditions, etc., as noted above, with appreciable frequency) (Bender et al., 2021; Weidinger et al., 2022).

Explainability is an essential ethical measure for mitigating these risks. Increasing the transparency of GenAI systems enables explainability. This means that when problematic outputs occur, stakeholders (like model developers and auditors) can trace those outputs back to specific model behaviors or certain biases in the training data. This tracing is what we consider model explainability, and it is obtained with good reasons (Lipton, 2018). For example, tools that help explain model behavior—attribution analysis, feature importance scoring, and prompt tracing—can help uncover whether a biased output is the result of a biased input representation or whether it reflects the model's tendency to amplify learned associations.

Ensuring that organizations are transparent permits corrective actions, such as system retraining, data refinement, or algorithmic adjustment. If AI systems function in a black-box manner, organizations subjected to accountability must provide clear, understandable, and—most importantly—ethical explanations for the actions they take and for the results those "intelligent" systems generate. If an organization is not capable of "explaining away" the results, then perhaps accountability is an impotent prescription for AI governance.

6.1.3 Regulatory Compliance

It is almost universally accepted that it is not only an ethical matter but also a legal matter for GenAI to be explainable. These tools are being applied in consequential areas such as hiring, credit scoring, healthcare, and law enforcement. Their declarations can and do turn on very important questions of rights and very for-profit business decisions that knock on the doors of individuals and change their lives.

In places where legislative frameworks govern such spaces, GenAI must express itself in a human-readable manner. The most significant legal framework that stresses the importance of clarity is the Partial Differential Equation of the EU (it is not a PDE but a legal document).

In a nutshell, the GDPR requires organizations to conduct an audit of the model (not all AI models can be audited), ensure that the model is interpretable (not all AI models are interpretable), and generate documentation (not all organizations document well, even in plain English) that outlines the inner workings and logic of the model. Complying with this law is a big deal because organizations can face serious fines and lawsuits if they do not. Open and clear model operations encourage responsible innovation. In conclusion, the clear functioning of GenAI systems extends beyond helping people to trust the systems and serving as a sort of ethical safeguard. This is necessary to comply with international standards of data protection and is key to ensuring that human rights are upheld in an AI-driven world.

6.1.4 Debugging and Model Enhancement

It is essential for developers to comprehend how a model creates outputs to identify errors, false correlations, and unexpected biases. They often use tools such as SHAP,

LIME, and integrated gradients to help visualize which features the model is paying attention to when generating the output (Ribeiro et al., 2016; Lundberg & Lee, 2017a). We found that these contexts are often missing when content generation is discussed in the literature.

6.1.5 User Acceptance and Uptake

For GenAI technologies to be widely used, people must understand the AI processes that yield specific results. This is especially crucial in knowledge-dependent fields such as education, journalism, and e-commerce, where seamless operation—requiring precision, transparency, and, above all, context—is provided by GenAI technologies (Amann et al., 2022).

In these sectors, making AI systems more user-friendly is essential for enhancing their trustworthiness. If you understand what an AI system has done well enough to explain it to someone else—which amounts to a kind of old-fashioned human intelligibility—you are much more likely to use it, and the tasks it is used for will probably require a good amount of human judgment and critical thinking skills. This is a way of saying that decisions made with the assistance of AI are still fundamentally decisions made by humans.

The fact that AI architects can explain what is going on in a comprehensible manner—and that our half-lives are much longer than those of the AIs we work with—means that a much wider swath of society can understand and thus influence the decisions that an AI can make. If XAI provides a depth of understanding and a kind of interaction that lets many relatively untrained people feel like they can really comprehend AI in operation, then AI is much more likely to be accepted—if not for any other reason than it looks less like something out of Arthur C. Clarke's "2001" and more like an actual, working thing. Figure 6.1 shows importance of explainability in GenAI.

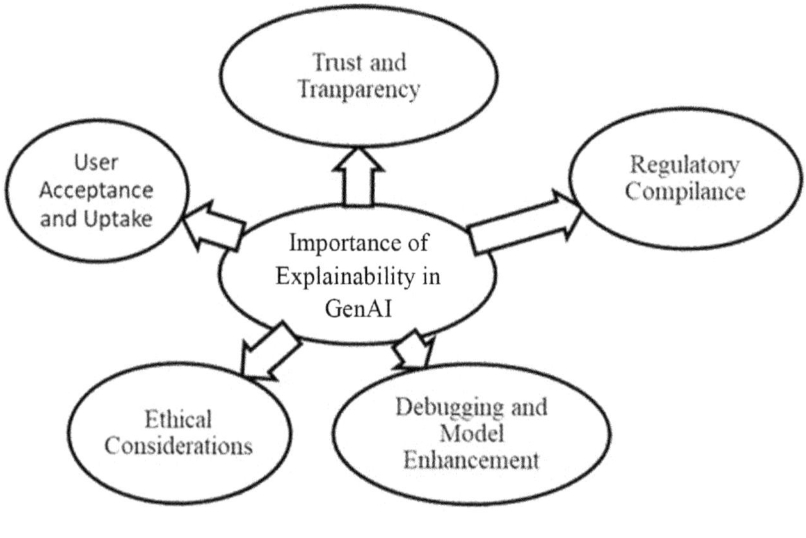

FIGURE 6.1 Importance of Explainability in GenAI

6.2 CHALLENGES IN EXPLAINING GENAI

Even though calls for transparency and accountability get louder, we cannot explain why GenAI works the way it does; that is a real challenge. The models themselves are extremely complex and tangled. Moreover, these AI tools—GANs, VAEs, and GPT-style transformers—are massive neural networks that have become huge data crunchers. There are millions or billions of tiny knobs; when set right, they create wild swirls; therefore, it is like a storyteller, but far better. There are several variables. Tracing an outcome back to a single decision is almost impossible.

The randomness in GenAI outputs amplifies this problem. Old-school AI gives the same answer each time you give it a query; generative models might give a different answer even when the input does not change. Some think it is a creative edge; therefore, it just adds extra trouble to the already hard job of matching results with the decisions that came first-A major problem, GenAI models remain hidden; one can never tell what they are doing inside. Most models do not provide a clear view of what is happening inside; therefore, you end up guessing. Therefore, as a developer or user calling 911 for GenAI, you want to see if it hides bias, gets risky, or acts unethically; you will find it hard because the model is anything but honest or clear. Therefore, there is no single test for XAI; everyone makes up their own rules, so comparing them is a nightmare. Often, we look at the percent right, hit–miss balance, and total error; therefore, those numbers alone do not prove that the model's answers are understandable or justified.

That thing gets tougher, because nobody has one standard everybody agrees on to say what counts as a "good explanation," especially when the models fire out answers. However, can this field continue to expand? Scalability turns into a serious roadblock. Saliency maps, feature attribution, and attention visualizations work well on small models, but when you move to giants like GPT or DALL·E, they hardly keep up. Models keep expanding, and they seem to fit more places; therefore, attempting to explain them instantly across dozens of decisions becomes a massive computational load. Therefore, explainability is closely linked to data bias and the ethical issues that arise from it.

GenAI is exposed to massive, almost never-cataloged data loads, which already bear historic and social bias; therefore, the bias seeps into the system. If we cannot explain it, finding those biases is barely possible, and fixing them across all the model's layers is practically impossible. Therefore, generative systems could end up producing content discriminatory, offensive, or simply wrong. This means using strong computer tricks, ways to peek inside them, and clear rules that make AI open, fair, and true to what we care about.

6.2.1 Complexity of Deep Neural Architectures

Current GenAI is based on deep neural networks. These mathematical models are composed of dozens, if not hundreds, of layers and millions or even billions of narrow settings that are almost impossible to see.

Therefore, it is virtually impossible to know how an input to the model morphs into an output. All the steps taken are like a giant math book that nobody can read.

Old rule-based programs had the virtue of being easily opened and inspected. Their explanations were plain to see. However, fancy math—math that is understood by only a few people in the world—has become the only viable way to explain these large, complicated models. Their fancy math appears to be a mess. As Rudin (2019) pointed out, this organizational messiness is arguably a bigger roadblock to real-world accountability than the sheer size of today's neural networks.

6.2.2 Trade-Off between Accuracy and Interpretability

Here is the classic situation: you want a model that not only performs well but is also interpretable. You sacrifice predictive power when you choose to work with a simple model, such as linear regression or a decision tree.

These models are easy to follow, and because their mechanisms are so straightforward, one understands why they make the predictions they do. When a model with many more parameters is used, interpretability is sacrificed.

Models such as the decision tree ensemble described in Section 3.7 of Chapter 3 are certainly more powerful, but they come at a cost when trying to understand them. State-of-the-art systems, such as GPT-3 or Stable Diffusion, serve up results that blow your mind, but their inner workings are a black box that you cannot peek into. In places where the cost of a mistake could be huge, such as hospitals and banks, it is better to understand what you are working with (Doshi-Velez & Kim, 2017). This classic trade-off is very real, and it has been around long enough that it is well understood by the machine learning community and others.

6.2.3 Domain-Specific Barriers (e.g., Healthcare versus Finance)

Various fields require a range of explanations. In the medical domain, practitioners require an unambiguous, medically grounded narrative as to why a model indicates a given treatment. The rationale must mesh seamlessly with the lore of the clinic—the knowledge and wisdom amassed by physicians over centuries.

Healthcare is not the only field where this sort of thing happens, but it is one of the best-developed examples. Similar dynamics—and the same principles of good teamwork and clear communication—also play out in law, finance, and all types of engineering. Indeed, there are often so many compliance and resource-allocation "rules of the road" such that keeping your headlights on becomes a full-time job.

6.2.4 Human Cognitive Limitations in Understanding Models

Even a flawless explanation can be too much for the intended audience to absorb. Miller (2019) indicates that people prefer concise, intuitive explanations and tend to disengage when faced with long-winded, detailed narratives. The art of explanation seems to hinge on telling a story that is both close to the actual phenomenon being explained and easy to understand. When faced with a heavy narrative, people are likely to misunderstand or ignore it. To make it even more challenging, we must consider that human interpretation can be quite biased. We generally see what we expect to see, and if people do not expect a narrative to be "good," they are less likely to absorb it. Thus, explanations must also contend with the problem of biased reception.

6.2.5 Risk of Oversimplification

In creating understandable explanations of their learned representations and decision-making processes, tools such as LIME and SHAP simplify the model's complexity. They attempt to take a complex model's representation and decision-making process and make it appear simple. This is beneficial because it helps people understand what is happening and what features are important for the model's decision.

However, simplifications can be dangerous. They can miss hidden biases or errors inside the model, and if an explanation is good enough that it could be mistakenly taken as the whole truth, it is potentially misleading. It is a difficult balance between making a model understandable and making it "look like" it is understandable, and I would argue that generative models have not yet mastered this balance.

6.3 METHODS OF EXPLAINABILITY IN GENAI

As GenAI models increase in size and complexity, gaining insight into their inner workings—that is, what makes them produce outputs—has become increasingly important. Their outputs should not be opaque; if they are, we may have nothing more than highly sophisticated black boxes. In the absence of clear guidance on how GenAI accomplishes what it does, we can only imagine (fatalistically or otherwise) whether it is really aligned with our goals, whether its outputs are potentially good or potentially bad, and whether its very existence is a threat to human artistic creativity. It is not sufficient to understand this only partially; we must have some grasp of the whole thing, or we are likely to make some very serious misjudgments about it. GenAI models could seriously harm us if we misjudge them, especially when their outputs are used in high-stakes applications.

6.3.1 Feature Visualization Techniques

In GenAI, techniques are commonly employed to gain insights into the model and how it processes input data. These methods "step inside" the model to determine which internal features are most significantly responsible for producing the output.

This type of probing is particularly illuminating when assessing model interpretability, as it provides key signals about which features matter and, more importantly, how they matter. Black-box models are among the best tools for assessing trustworthiness (Montavon et al., 2018).

6.3.1.1 Activation Maximization

The main purpose of Activation Maximization is to determine the inputs that cause particular neurons or components of a model to activate the most. This method requires the creation of a type of input data that contributes to the optimum stimulation of a specific neuron. After having some understanding of the kind of thing that the network recognizes, researchers can then work on discerning the features (the types of "good stimuli") that the network uses as the basis for its recognition performance.

6.3.1.2 Saliency Maps

Saliency maps offer a more focused way to interpret models, showing the exact spots in the input that are most responsible for their predictions. For tasks such as image generation or classification, these maps can be used to determine which pixels are the most vital in the decision-making process. Language models are synonymous with this process when determining which words in a sentence carry the most weight. The use of saliency maps is a direct way of not only seeing what parts of the input are most important but also assessing whether the model is attuning to the right features to get the right answer.

6.3.1.3 Latent Space Exploration

Latent space exploration focuses on models like VAEs and GANs, which convert data into representations in latent spaces. Researchers can probe the semantic understanding of a model by moving through or interpolating the latent space. They have examined how tiny modifications to the latent vectors result in large shifts in the model outputs. This provides insight into the structure of the latent space and, by extension, the type of space that our models have learned to understand. All of this probing is, of course, a good thing. After all, we want to be confident that the models we build are working in a morally unexceptionable manner—obtaining features and making decisions in ways that we can understand and that we can be sure will not lead to harmful outcomes.

6.3.2 Model-Agnostic Techniques

There are various ways of making ML models explainable, and the methods considered here have the virtue of being model-agnostic. In other words, they do not depend on the specific structure of the model, such as decision trees or neural networks, that is being scrutinized and thus can be applied to any number of generative systems. One such technique is SHAP, which allocates significance scores to features in a manner reminiscent of game theory. In the way that games can be understood in terms of the payoffs won by players, SHAP reveals the extent to which features affect predictions.

A complementary technique is LIME, which builds very simple and understandable models, such as linear regression, which are good approximations of the more complex model being explained, but only in the region around a certain prediction. A very different and somewhat newer method for producing explanations is based on counterfactuals. Counterfactuals are essentially "what if" scenarios, and by answering them, one can generate a kind of explanation. Why did the model output A instead of B?

Layer-wise Relevance Propagation (LRP) provides a way to understand how a model makes decisions. It does this by working backward through the layers of the model. Relevance scores are assigned to the input features, which helps in understanding the decision at the global level. Just as with any explanatory method, there are noteworthy pros and cons to using the LRP.

These scores clarify the contribution of each input feature to the final output. They provide a much clearer picture of model behavior, especially in the case of image and text generation tasks, for which LRP was primarily designed. In these tasks, it is essential to know which part of the input the model is working with.

XGANs, or explainable generative adversarial networks, address the opacity problem that is usually associated with standard GANs. They aimed to make GANs more interpretable, which in turn should make the outputs of a GAN more understandable. XGANs achieve this by adding a conditional component to the GAN.

Distinct semantic attributes (e.g., color and object shape) are assigned to separate latent dimensions in disentangled representations. This allows the generated content to be controlled and explained. Encouraging a GAN to produce outputs that closely resemble real instances or prototypes is what we term as using prototype-based explanations. This is a powerful method for linking outputs to meaningful real-world instances.

A GAN attribution technique, such as Grad-CAM, cannot be applied directly to old GAN architectures, such as the first Pix2Pix or CycleGAN. Conventional GANs generate images at the pixel level, meaning that each pixel is treated individually.

Regarding attribution techniques, old GANs are akin to traditional neural networks that only output a task at hand, for example, a classification task. However, if one were to look closely at a traditional neural network, one would find that techniques such as backpropagation for gradients and guided back propagation exist.

Models such as Transformers utilize attention mechanisms that are inherently transparent. They show which parts of the input are most important during the generation process. When generating text, some words are more influential than others in determining the output of the text. Attention scores reveal this type of ranking. In contrast, when generating captions for images, attention can also be understood in a spatial sense: attention maps indicate the relevant regions of an image when producing a caption.

The explainability of Interpretable Architectures means building models that are transparent in and of them. We define the integration of AI and Neuro-AI into Symbolic AI as follows:

Neural networks are good at generalization. They can learn from massive amounts of data, identify patterns in these data, and make predictions about future data based on the past data on which they were trained. They work well when the dimensions of their input space are extremely high and the number of training examples is very large. Networks are good for approximating nonlinear functions. However, on the other hand, networks are not very transparent. They effectively encode a hidden representation of the learned knowledge but do not provide human-readable, interpretable, or understandable logic or reasoning.

This toolkit, which never stops evolving, helps researchers and developers comprehend the far more intricate behaviors that GenAI systems now display. To narrow the gulf between performance and transparency, we must carry out certain actions,

TABLE 6.1
Summary Table of Explainability Methods

Method	Description	Example Techniques
Feature Visualization	Identifies how inputs activate specific neurons and what structures the model learns.	Activation maximization, saliency maps, latent space exploration
Model-Agnostic Techniques	Works with any model type to explain how input features affect outputs.	SHAP, LIME, counterfactual explanations
Layer-wise Relevance Propagation (LRP)	Backpropagates relevance scores from outputs to inputs.	Layer relevance scoring for image/text generation
Explainable GANs (XGANs)	Adds transparency to GANs by disentangling features or using prototypes.	Disentangled representations, prototype matching, attribution techniques
Attention Mechanisms	Shows which parts of the input are most relevant during generation.	Attention maps in Transformers, attention scores in NLP
Interpretable Architectures	Builds models with inherent transparency or modularity.	Neuro-symbolic AI, modular generative systems

especially audits, to ensure that these models remain trustworthy and aligned with human values. Table 6.1 presents all methods of explainability.

6.3.3 Intrinsic Explainability Approaches

The intrinsic method of explainability seeks to construct an interpretable model from the start. Classic interpretable models include decision trees, sparse linear models, and attention-based networks. In these cases, humans can examine the model's parameters and see something meaningful. An increasingly popular approach for creating interpretable models is to use a clear set of latent variables. These hidden variables have been forced to take on comprehensible, human-meaningful values (as in the above example). Even when the variable values are encoded in a more complex manner, interpretability constraints can still be imposed on the model.

For example, we can build symbolic reasoning layers into a neural network. These layers are good at thinking through a problem and can generate a simple plan for solving the problem. After the net has made a basic plan, it can use the reasoning layer to fit the observed data to that plan. Both of these steps—planning and data fitting—are now more interpretable because we use a neural network whose low- and high-level parts are understandable.

6.3.4 Counterfactual Explanations

Counterfactual explanations inquire, "What minuscule alteration to the input would reverse the output?" They present a clear alternative that aligns with human conceptions of causality. In the context of text generation, a counterfactual would be "If you were to substitute the word 'awesome' for 'average,' the model's tone would shift from promotional to neutral."

For image synthesis, one could use background lighting to analogize how a different cause could produce a different effect—that is, how a human might interpret a change in appearance if the "right" image were to be generated. These are good because they make the model's boundary visible and show its sensitivity to certain features. This intuitiveness helps users check whether the model behaves as expected.

However, creating counterfactuals for large GenAI systems requires a significant amount of computing power. We must search a huge space for the smallest change, especially when the models have thousands of hidden dimensions. Sometimes, the answer the system provides is unrealistic, such as suggesting a medical dose that does not exist. In these cases, the explanation, while correct in math, is not useful for a human. Despite the cost, counterfactuals bridge the gap between a black box and human reasoning, allowing us to examine the model.

6.3.5 Example-Based Explanations (Prototypes and Criticisms)

Example-based methods show how a model works by showing *prototypes* (typical cases) and *criticisms* (edge or failure cases). For a text generator, a prototype could be a sentence that resembles the model's average style. A criticism might be a bizarre or biased sentence that the model sometimes produces. Studies have shown that concrete examples help people learn what models do.

These techniques also reveal the influence of the training data. Prototypes usually reflect the most prevalent patterns, whereas the most common criticisms target outliers. However, two issues emerge. First, it is difficult to select representative examples. An analyst can make an example-based method look good by selecting only cases that are easy to understand and hiding the places where the model lacks trustworthiness. The second issue is that if we want to base trust on the identification of fair and clear selection of prototypes and criticisms, it requires a ton of computation.

6.3.6 Natural Language Explanations and Narrative-Based XAI

Explanations for the model's reasoning were transformed into natural language, rendering them into sentences that a human could read and understand. An explanation might say: "The model looked at historical sales data and gave more weight than usual to feature X for this prediction." This step is useful for humans without access to the model's internal workings. They could read these explanations and understand the model's outputs better than they would if we presented the outputs in some other way. However, nowadays, when we talk about AI explanation, we are really talking about whether LMs can faithfully do what they are built to do; that is, whether they can give us faithful, understandable, human-level sentences that tell us how or why the model did a thing.

Therefore, we need to measure how well the story matches the actual computation. The balance between clarity and truth remains an open research question. Overall, narrative-based XAI has promise for making generative AI transparent but only if we put checks in place to stop the system from deceiving us.

6.3.6.1 Synthesis and Outlook

The review above shows that no single explainability trick works for everything in the context of generative AI. Post-hoc tools, such as LIME and SHAP, provide fast, model-agnostic clues but can be fuzzy; intrinsic designs provide clear inner workings but may lose power; counterfactuals and example-based methods are human-friendly but can be costly; and natural language stories are easy to read but risk being untrue.

In high-risk areas—legal arguments, medical advice, school grading—trust, accountability, and rules demand solid, multilayered explanations. Therefore, researchers should aim for hybrid systems that combine the strengths of several methods, adapting the plan to the risk level of each use.

Many research gaps remain: making post-hoc tricks more accurate, letting intrinsic models stay strong, finding realistic, cheap counterfactuals for huge models, speeding up prototype and criticism search, and keeping natural language reasons honest. Solving these problems will be key to allowing generative AI to move into places where clear thinking is not optional but required.

6.4 RESEARCH ASPECTS IN EXPLAINABILITY FOR GENAI

Explainability in GenAI is now a huge focus, as the models become messier, stronger, and everywhere. Because the core sits at the center, people launch massive—some even call them futile—efforts to address the many problems GenAI and its offsprings bring; therefore, the push seems never-ending. Therefore, this study attempts to make GenAI systems easier to understand, more trustworthy, and aligned with the ethical standards we humans hold. This big research push building global explainability scores (XGMs) so that we can see how clear an AI's choices are. There is no single way to grade how easy it is to understand a model; therefore, attempting to compare what is simple for one type of model versus another is complicated. For example, a decision tree could be replaced by another type, such as a random forest, which would shift the method (e.g., something that works like a neural network). Therefore, the researchers try to devise plain, steady guidelines that could fit almost any architecture; does that actually work? Therefore, AI tools such as GANs, VAEs, and Transformers can create pictures, spin text, and even output videos. This is what we obtain when we benchmark models, and it helps us check whether they work correctly. Therefore, a major research aim is trying to build generative models that feel fair, it also tries to reduce bias—doesn't that sound right? They learn from huge piles of online forums, news feeds, and social posts that hide bias; therefore, we must ensure that they do not repeat old stereotypes or create fresh unfair results; ideally, they act more responsibly than most human replies. They are testing ways to strip bias, draft moral checklists, and shape training to stay fair; therefore, the research keeps moving forward. Researchers have tested hybrid methods that combine symbolic reasoning and deep learning.

Therefore, they hope that it will reveal new patterns. Although symbolic AI has long offered rule-based thinking, deep learning has now taken over, finding endless patterns, and powering almost all the fast AI apps and smartphones we own. How is that possible? Adding these two halves, the hybrid model we studied is

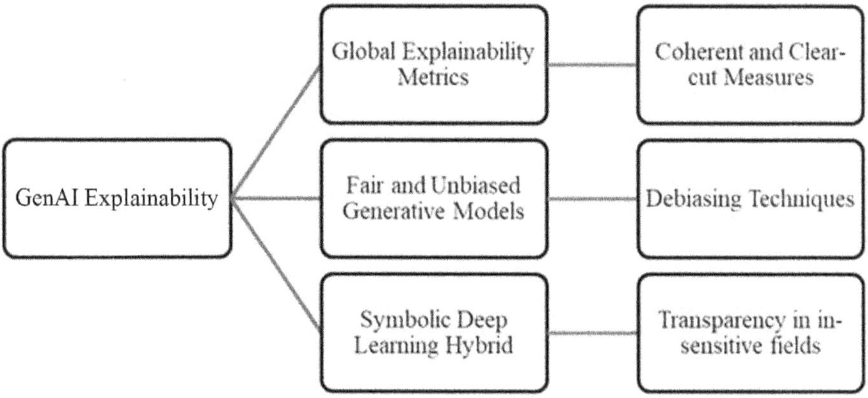

FIGURE 6.2 Flowchart for GenAI Explainability

understandable, yet it still performs well. Since it mixes everything together, it is a real win for any area that needs plain truth; therefore, courts and health care benefit. Another aspect of multimodal GenAI is the explainability of multimodal issue. In this case, the model not only spins out answers but also grabs inputs of several kinds, all at the same time. **Therefore, which output types could work**—text, a picture, a video? Explainable multimodal AI is more complex than plain XAI, as multimodal AI must process a tangled mix of information from pictures, sounds, and videos. In short, the work on GenAI explainability moves fast; therefore, the aim is to have models that are strong, easy to obtain, accountable, and fair. Rules like the GDPR want the text to be easy to read, and they say that AI must be used the right way; therefore, there are no shortcuts. Research on GenAI explainability is still needed; this is what ensures that new technology remains responsible and that people choose to follow the rules on their own. Therefore, the ethical use of GenAI in real life hinges on people choosing to follow the rules. Flowchart for GenAI explainability is given in Figure 6.2.

EXERCISES

1. What makes explainability so important in the area of GenAI for different stakeholders like developers, regulators, and end users?
2. How does the lack of clear explanation impact the reliability and acceptance of GenAI systems?
3. What role does explainability play in ensuring ethical and accountable AI deployment?
4. In what ways do the requirements for explainability differ among various sectors like healthcare, finance, and education?
5. What are the real-world consequences of deploying GenAI systems that lack explainability?
6. How do model-agnostic techniques such as SHAP and LIME operate to clarify AI predictions?

7. In what ways can LRP be tailored for generative models?
8. How are attention maps used to interpret transformer-based generative models?
9. What are the limitations of employing SHAP and LIME for high-dimensional generative output?
10. How effective are these methods in rendering generative models transparent to non-technical users?
11. What are XGANs, and how do they differ from traditional GAN architectures?
12. What techniques are available to enhance the interpretability of latent spaces in GANs? ·
13. How can interpretable models be integrated with GenAI without compromising their performance? ·
14. What advantages do disentangled representations offer for enhancing interpretability? ·
15. Are there any trade-offs between the explainability and creativity of the outputs of generative models?
16. Why are GenAI models frequently regarded as black-box systems?
17. How does the stochastic nature of their outputs complicate efforts to explain them?
18. What challenges emerge when scaling explainability tools for large generative models, such as GPT or Stable Diffusion?
19. What research directions are being pursued to enhance the explainability of GenAI on a large scale?
20. How can explainability be balanced with computational efficiency and model performance?

REFERENCES

Amann, J., Vetter, D., Blomberg, S. N. F., Christensen, H. C., Coffee, M., Gerke, S., & Schönberger, D. (2022). Explainability for artificial intelligence in healthcare: A multidisciplinary perspective. *BMC Medical Informatics and Decision Making*, 22(1), 1–9.

Bender, E. M., Gebru, T., McMillan-Major, A., & Shmitchell, S. (2021). On the dangers of stochastic parrots: Can language models be too big? *Proceedings of the 2021 ACM Conference on Fairness, Accountability, and Transparency (FAccT).*

Davenport, T. H., & Ronanki, R. (2018). Artificial intelligence for the real world. *Harvard Business Review*, 96(1), 108–116.

Doshi-Velez, F., & Kim, B. (2017). Towards a rigorous science of interpretable machine learning. *arXiv preprint arXiv:1702.08608.*

Gilpin, L. H., Bau, D., Yuan, B. Z., Bajwa, A., Specter, M., & Kagal, L. (2018). Explaining explanations: An overview of interpretability of machine learning. *2018 IEEE 5th International Conference on Data Science and Advanced Analytics (DSAA)*, 80–89. IEEE.

Goodman, B., & Flaxman, S. (2017). European Union regulations on algorithmic decision-making and a "right to explanation". *AI Magazine*, 38(3), 50–57.

Guidotti, R., Monreale, A., Ruggieri, S., Turini, F., Giannotti, F., & Pedreschi, D. (2018). A survey of methods for explaining black box models. *ACM Computing Surveys (CSUR)*, 51(5), 1–42.

Lipton, Z. C. (2018). The mythos of model interpretability. *Communications of the ACM*, 61(10), 36–43. https://doi.org/10.1145/3233231

Lundberg, S. M., & Lee, S.-I. (2017b). A unified approach to interpreting model predictions. *Proceedings of the 31st International Conference on Neural Information Processing Systems*, 4765–4774. NIPS.

Miller, T. (2019). Explanation in artificial intelligence: Insights from the social sciences. *Artificial Intelligence*, 267, 1–38.

Montavon, G., Samek, W., & Müller, K.-R. (2018). Methods for interpreting and understanding deep neural networks. *Digital Signal Processing*, 73, 1–15.

Radford, A., Metz, L., & Chintala, S. (2016). *Unsupervised representation learning with deep convolutional generative adversarial networks*. arXiv preprint arXiv:1511.06434.

Ribeiro, M. T., Singh, S., & Guestrin, C. (2016). "Why should I trust you?": Explaining the predictions of any classifier. *Proceedings of the 22nd ACM SIGKDD International Conference on Knowledge Discovery and Data Mining*, 1135–1144. https://doi.org/10.1145/2939672.2939778

Rudin, C. (2019). Stop explaining black box machine learning models for high stakes decisions and use interpretable models instead. *Nature Machine Intelligence*, 1(5), 206–215.

Samek, W., Wiegand, T., & Müller, K. R. (2017). Explainable artificial intelligence: Understanding, visualizing and interpreting deep learning models. *arXiv preprint arXiv:1708.08296.*

Wachter, S., Mittelstadt, B., & Floridi, L. (2017). Why a right to explanation of automated decision-making does not exist in the General Data Protection Regulation. *International Data Privacy Law*, 7(2), 76–99.

Weidinger, L., Mellor, J., Rauh, M., Griffin, C., Uesato, J., Huang, P.-S., . . . & Gabriel, I. (2022). Ethical and social risks of harm from language models. *arXiv preprint arXiv:2112.04359.*

Weller, A. (2019). Transparency: Motivations and challenges. In *Explainable AI: Interpreting, explaining and visualizing deep learning* (pp. 23–40). Springer.

7 Bridging Fairness and Explainability in AI

AI systems are being used increasingly in places such as finance, healthcare, and hiring. In these spots, the results can be huge—sometimes it means a life or a job. Therefore, two ideas are important: fairness and explainability. Fairness means that the program should treat everyone the same and not favor one group over another. Explainability allows people to understand why a machine makes a certain choice.

These two notions are intertwined. Complicated models are difficult to check for bias. If we cannot determine whether a model's choices are fair, we cannot call it trustworthy. If we cannot understand the model well enough to explain its reasoning, how can we hold it (and the people who created it) accountable? How can we have any confidence that, to the extent it really is a model of intelligence, it is as good a model of something like fair intelligence as a human is expected to be? Old-school statistics have some ideas for making such models clearer. Some modern "black-box" methods are, at best, murky, and tend to mix with our old ideas. What about mixing fresh methods with our classic ideas and coming up with a "smarter," fairer, and more understandable AI? Hearing the key ideas from the old and new could lead to a better mix in achieving a clearer line of sight to model transparency. Table 7.1 gives key concepts for fairness and explainability in AI.

7.1 THE INTERPLAY BETWEEN BIAS AND OPACITY

Many AI applications create the illusion that they perform everyday tasks much better than humans. Consider our picture-recognition software. This app, by all appearances, has a kind of visual supremacy when, in fact, it has none. It can identify the kinds of things you want it to identify—both common and really out-there kinds of things—with a level of accuracy that impresses most people. However, if you and I were to have a face-off with our own visual systems, I really do not think the app would win—and this is because of a reason that gets us only so far in rethinking its kind of supremacy. For software program, understanding the way we do is not possible, and that is not seeing when it comes to those dual systems.

Therefore, this technology might help, although it is still far from flawless in many real-world applications. Predictive systems, from spam blockers to stock suggestions, appear to shape decisions. However, these strengths may mean that we trust them too much, and their flaws may remain hidden. Therefore, we should monitor our reliance on such machines. Figure 7.1 shows interplay between fairness and explainability.

One major reason DL feels mysterious is the way its networks are built. They use many layers of tiny artificial neurons that learn to extract information from data that

DOI: 10.1201/9781003713951-7

TABLE 7.1
Key Concepts of Fairness and Explainability in AI

Concept	Description	Tools/Techniques
Fairness	Ensuring unbiased and equitable AI decisions across demographic groups	Fairness metrics, re-weighting, adversarial debiasing
Explainability	Making AI models' decisions understandable to humans	SHAP, LIME, saliency maps, counterfactuals
Interplay	Knowing how explainability supports identifying, measuring, and fixing bias	Fairness audits, human-in-the-loop, regulation

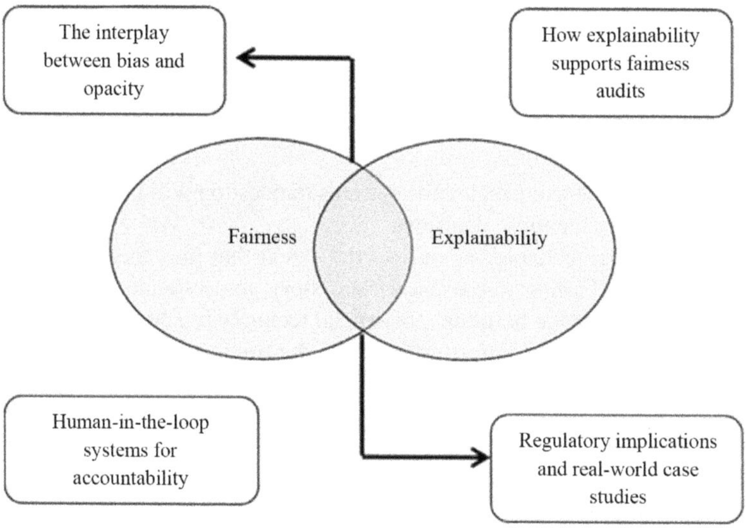

FIGURE 7.1 Interplay Between Fairness and Explainability

become increasingly vague as you go deeper. Consider a face recognition program. The first layer might spot simple edges or color patches.

Later layers start to group those bits together and finally try to identify the actual person in the photo. However, the "features" the system uses are not written by anyone. They are simply patterns that the model grabbed from a huge training set. This is where opacity may occur. When we see the final answer and try to trace it back to the bits that led to it, the path is almost invisible to us. Traditional rule-based or linear models work differently; they provide a clear chain of reasons.

This lack of clarity can cause several problems. Bias may hide inside the model if the data it learned from were already unfair. The system could then repeat or even boost these unfair patterns without our noticing. Accountability is uncertain. If an AI says no to a loan that should be approved or mislabels a health problem, it is very

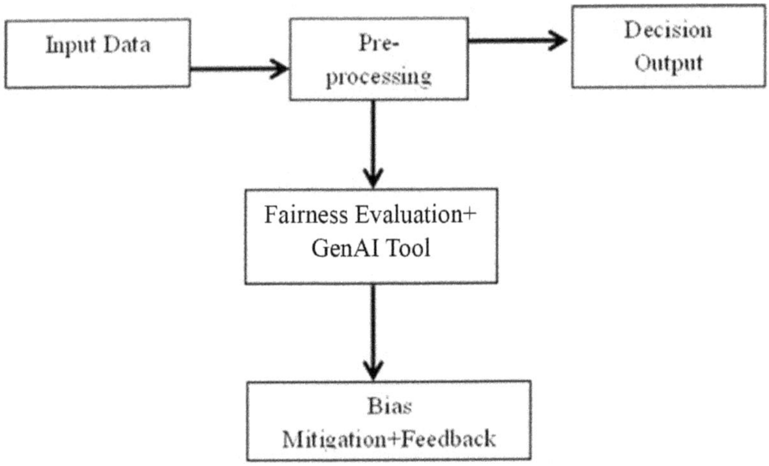

FIGURE 7.2 Bias and Mitigation Process

likely that the humans who created the systems in question will be unable to explain why the machines made those decisions.

Most people trust technologies only to the extent that they feel they understand how they function. Failing to present a clear story, understandable to the average human, can drive a wedge between a beneficial technology and the society it would like to help. This would be unfortunate. It may be time to ask whether researchers can or should do what the manufacturers of these technologies certainly should do: tell stories. Figure 7.2 shows the bias and mitigation process.

- Hidden bias can occur when things like race or gender sway how a model decides, and if we cannot see why, the system might keep unfair patterns alive (Barocas et al., 2019).
- Lack of accountability occurs because people cannot be blamed for outcomes they do not understand (Doshi-Velez & Kim, 2017).
- Ethical concerns arise when choices hurt marginalized groups, and no one calls them out (Mehrabi et al., 2021).

These problems usually begin with data. The data may be skewed, dirty, or limited. Bad feature picks or missing examples can push a model to be incorrect. The result is a loan could be denied to a low-income applicant for no clear reason. A patient may receive a wrong diagnosis. Therefore, the need for a clear explanation seems obvious, yet difficult to achieve.

7.2 EXPLAINABILITY AS A TOOL FOR FAIRNESS AUDITS

Ensuring the clarity and openness of models appears to be essential for their safe use. Developers, regulators, and end users must have access to understandable accounts

TABLE 7.2
Example Outputs from SHAP Analysis in a Hiring Model

Resume ID	Gender	SHAP Importance for "Gender" Feature	Predicted Hire Probability	Counterfactual Change
001	Male	0.02	0.86	—
002	Female	0.12	0.47	↑ to 0.85 (if gender changed to male)
003	Female	0.09	0.52	↑ to 0.84

of model behavior if they are to assume responsibility for that behavior. Accounts of behavior are most illuminating when they are tied to model features in ways that are not just obvious post hoc but also evident a priori to someone reading the account. Identifying the features that matter may help find bias or unfair outcomes that the model could show. Options such as SHAP, LIME, and counterfactuals often appear in practice.

Personally, we think the key point is that getting a sense of what drives the model, especially its priorities, appears to be needed for trust. Nevertheless, the process has limits and may never capture every factor. Table 7.2 shows some output from SHAP.

7.2.1 Feature Importance Analysis

Looking at which features matter is the first step in making AI clear. This may seem simple, but it helps us understand what a model is doing. The idea is to determine how much each input contributes to the final answer. ML models often derive insights using two widely used tools: SHAP and LIME. Both do profoundly similar things for different reasons: they provide human-understandable interpretability for the otherwise inscrutable outputs of complicated models. SHAP tends to show a more complete global view of the model, whereas LIME provides a clearer picture of what an individual prediction (or instance) is doing. Intuitively, one would think that one or the other would be sufficient for model evaluation.

However, these two tools help illuminate different aspects of the model's workings, so they are kind of like "cheating" for AI ethics. If the model is fundamentally creating an output for the wrong reasons, then it will hit a "wall of badness" when the veer-off-path reaches the important variables.

7.2.2 Counterfactual Explanations

Delving into the black-box nature of ML models can be achieved using counterfactual explanations. These involve posing simple "what if" questions, which, for the most part, even a child could understand. Consider the scenario where we alter a loan applicant's gender from male to female, while leaving all other factors unchanged. By altering only one input and keeping all other variables stable, we may observe a significant change in the model's output (Wachter et al., 2017). When a small change in the input causes a huge flip in the output, it suggests the presence of some big problem, such as bias. This raises questions regarding our confidence in these models. We need to have confidence in them to use them for bigger and better things. However, when we look at them under the lens of counterfactuals, it seems like we have some underlying issues that we have to use tunnel vision to look past. Shifting the focus, counterfactuals can also be used to identify hidden discrimination in models. These tools are powerful but can be used irresponsibly (Barocas et al., 2019).

7.2.3 Algorithmic Debiasing

When an AI system makes a biased decision, we often try to correct the bias and understand why the system makes a discriminatory choice. One possible solution is to add an adversarial network to the biased model so that the new system ignores race and gender in decision-making. Although this method does not guarantee perfect fairness, it may push the model toward a more even playing field (Hardt et al, 2016a). Another potential solution is to reweight the model inputs.

This pushes the model to be more reliant on things like "does the person have a job?" instead of relying on hidden proxies that act as race, gender, or socioeconomic status. Both of these potential solutions could help improve the fairness of the system without making it comprehensible to humans. However, why is this important? After all, a biased AI might look like a biased human, and if we cannot understand the AI, we cannot even say with confidence that it is better than a human who makes the same biased decision. Researchers have proposed testing it on many types of groups (Kamiran & Calders, 2012). However, even with this recommendation, the results can vary significantly. What if the "it" in question has hidden biases? Should we then test these biases in a controlled experiment before going public with the findings?

7.2.4 Regulatory Implications

Explainability in AI is no longer just a technological bonus. This may mean a legal and moral obligation. AI is present in many aspects of human life. AI is used in credit scoring, hiring, healthcare, and even policing to decide big things. Therefore, regulators worldwide are beginning to ask for more openness and responsibility from the makers of these systems. One major rule is the EU's GDPR, which started in 2018. It is about machines deciding things for us and building profiles of people. Therefore, you have a right; you do not have to let a lone machine's decision call the shots.

Consequently, any decision capable of impacting an individual's life circumstances or legal standing assumes critical significance. As for the law, it mandates that you have the right to request clear and meaningful explanations about how the reasoning functions.

Therefore, some explainability is needed for high-risk AI. So the next EU AI Act? It is basically built on the GDPR, which extends the same rules to AI. It splits AI into risk groups, allowing for the rapid ones fast. High-risk tools (such as those that select jobs or grant credit): strict rules on openness, human oversight, and proper paperwork required. The Act: It requires proof that the system is safe and works correctly; therefore, the proof must remain valid while the system runs. Please provide the full sentence that you would like to be rewritten. In 2017, the findings revealed a clear pattern of behavior (Madaio et al., 2022).

Therefore, proof should show how data are collected, how choices are reached, and what steps are taken to cut bias and maintain fairness. **The clarity with which you have described the system is crucial to all those factors.** Some think it might hold back new ideas; therefore, trust matters more, and the delay might be acceptable. It cannot provide clear explanations; therefore, AI risks losing public confidence. Both universities as well as firms are stuck trying to meet the rules. Both tools and training are essential; without them, nothing functions effectively. Otherwise, the consequences of failing to comply could become problematic.

Explainability is important when firms try to follow AI rules; therefore, they can see why a decision occurred. Isn't that clear? First, the system should provide a straightforward reason for each answer, plain enough, with no guessing. When a loan, job, or even a medical test is turned down, people want to know the reason, as without it they feel stuck. Is there a clear and simple explanation? This could be what pushes them to ask for a review. Second, many regulators want transparency reports and clear information. Therefore, these reports explain how the algorithm runs. They name the students in low-income neighborhoods as it lifts or pushes down. They also set the fairness caps that they think are fair. They also stated which explainability method was used. Most reports look legal, but could they still hide bias that you cannot see? If a machine's choice hurts a person, explainability tools can chase the steps that led there (Chawla et al., 2002).

Therefore, the person harmed can claim that the result was wrong, even unfair; it may even feel unjust. Does this match the due process? Therefore, it could mean that the maker ends up being answerable. If a company skips these steps, it could be fined, face a lawsuit, receive bad press, or even receive a court order demanding the system be shut down. Adding a clear reason for AI decisions? Not just a bonus, it actually shields the business and earns people's trust; therefore, explainability is a must for any AI rule book. As rules become stricter everywhere, the need for AI that is clear, ethical, and socially acceptable continues to grow. Legal duties and Moral responsibilities: Public acceptance—therefore, it is all touched on. Therefore, firms should invest in simple tools that allow users to see why each auto-choice occurred.

7.2.5 Human-in-the-Loop Systems for Accountability

When AI is given large tasks to perform, it is crucial that humans are also involved in the process. If there were no humans in the equation, we could get into the mess we are already in, with poor decisions being made by AI. Maybe "no oversight = big problems" is the first lesson we teach in "Artificial Intelligence with Accountability." Humans should oversee the line calls made by AIs. If the machines are not making calls that can have serious consequences, then what is the point? When is the decision big enough that a human must approve it? A distinguished engineer named Allen Neuman says, "If a decision could make someone onto a bad path, then a human should oversee that decision."

Context is another area where humans have the advantage. Let's not forget, AI is pretty much dumb. Despite all the shiny talk about intelligence in AI, these entities

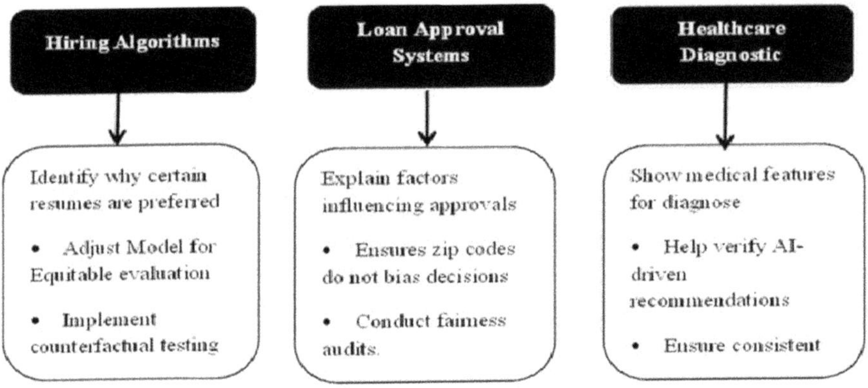

FIGURE 7.3 Fairness and Explainability

are just very successful parrots—"look up, look down, wait your turn to talk, and always try to sound smart and like the top half of the class."

How, then, does a human outperform? Well, without going too deeply into this, I would say it has everything to do with the historical intelligence of humans, who create and understand context within which intelligence functions (Miller, 2019).

The critical outcome of HITL systems is bias mitigation. Explainability tools, like SHAP or LIME, help human users understand why an AI system made a certain decision. When the output from a model appears to be biased or otherwise unjust, the human in the loop can override the model's decision and make adjustments. Those adjustments, and the rationale behind them, can then be used to inform the future retraining of the model itself (Binns, 2018).

Healthcare is an example of this. AI can parse several medical images and suggest a diagnosis. However, a human health professional—most likely a physician—will have the final say on the model's accuracy and will consider the fairness of its predictions across various demographic groups. Why does this matter? Because it is not enough that the model is accurate and interpretable, it also needs to be good enough to base life-and-death decisions on. Figure 7.3 summarizes fairness and explainability.

7.3 SOCIO-TECHNICAL PERSPECTIVES ON FAIR AND EXPLAINABLE AI

It requires far more than good intentions and human judgment to guarantee fairness and create intelligible AI systems. Avoiding these pitfalls definitely involves work that is more technical, better-designed, and better-engineered; that entails something even more consequential: the use and choice of data, the crafting of algorithms, and—most importantly—the humans who design these systems (Selbst et al., 2019).

A major source of problems with AI is the use of algorithms that humans have made unfair decisions when designing them or using biased data in the ML process. Ensuring that decisions are fair and that systems are interpretable is supposed to involve an assurance from the "human" side of the equation that they will not be used for unfair price discrimination. If we are not doing that, then what are we doing?

We should also worry about the negative impact hidden conversations could have when we assume we are being transparent, and how easily we could pit different communities against each other in the name of explainable AI (Ehsan et al., 2021). Consequently, this alteration implies that equitable AI is a social project almost as much as a technical one, not merely a matter of fixing the code. Vigilant ongoing community monitoring is required (Morley et al., 2021).

7.4 TECHNICAL METHODS BRIDGING FAIRNESS AND EXPLAINABILITY

AI systems become increasingly complex each year. Many researchers claim that fairness and explainability are urgent. Fairness implies nondiscrimination against any person or group. Explainability allows users to understand the rationale behind a decision, which may increase their trust in it.

Aligning both ideas is very difficult because methods must cut bias while also spelling out reasoning. Some recent tricks appear to bridge this gap, making the job easier. These approaches may lead to more ethical AI designs. However, doubts remain regarding the extent to which bias can be eliminated. Overall, progress points to fairer and clearer systems today (Amini et al., 2019).

- Adjusting training to make AI fairer
- Rebalancing data to mitigate bias
- Post-hoc analyses that make decisions understandable
- Bringing fairness and explainability into alignment

7.4.1 Adversarial Debiasing

Adversarial debiasing attempts to clean up the bias inside a model's learned features. This is achieved by adding an extra opponent network to the training loop. The main model still focuses on its primary job, such as sorting images into the correct categories. Meanwhile, the opponent learns to guess things that we do not want the model to use, such as a person's gender or race, from the same internal signals.

By pulling the two goals apart, the system is pushed to hide sensitive cues. This may mean that the classifier becomes fairer without affecting the overall accuracy. It could also slightly change the overall speed.

This technique ensures that the end model arranges things so that decisions are made without regard to protected information. This method contributes directly to fairer outcomes. Explainability techniques, such as SHAP or influence functions, can then be applied to show that because the final model does not use the protected information, it does not affect outcomes in an unfair manner.

7.4.2 Re-Weighting and Re-Sampling

Two data-level approaches to address class or group imbalances in the training dataset are re-weighting and re-sampling.

Re-weighting means assigning different weights to instances during training so that the model respects underrepresented groups. We do not want the model to be too respectful of run-of-the-mill cases and consequently too rude to instances that are not typical. However, we must be cautious when doing this.

Re-sampling includes oversampling examples of the minority class or undersampling examples of the majority class to balance the dataset. Models can be guided to learn fairer patterns using these methods. When used with explanation tools such as LIME or SHAP, stakeholders can see how model outputs are affected when features are more equitably distributed across populations.

7.4.3 Post-Hoc Explanation Tools for Bias Diagnosis

Methods for generating post-hoc explanations of model decisions, such as SHAP (Lundberg & Lee, 2017a), LIME (Ribeiro et al., 2016), and saliency maps, are therefore invaluable for understanding the outputs of black-box models. If you cannot understand the model, you certainly cannot trust it.

SHAP assigns an importance value to each feature for a given prediction based on the rules of game theory. LIME forms local proxy models that act like the black-box model in the neighborhood of a selected instance. Visualizing the most influential decision component maps, such as saliency. For decisions based on raw input images, such as classifying a cat, input components such as image pixels can be mapped to decision paths. The most influential image pixels have high saliency. These tools can pinpoint when a model is using sensitive attributes or their proxies in a big way (e.g., when a zip code is used as a proxy for ethnicity), and thus diagnose when bias might be a problem.

7.4.4 Explainable Fairness-Aware Training Pipelines

We adopted a comprehensive approach, emphasizing its integrated and holistic nature. When we say "integrated," we mean that it is built around both the principles of fairness and the ideals of transparency, and it suffuses the entire AI model lifecycle from the initial conception of the model to its real-world deployment.

The reason we pursue this integrated principle is crucial, especially in high-stakes domains such as healthcare and the criminal justice system. At the core of this integrated thrust is development not just with fairness in mind but with "fairness objectives." The latter is a somewhat arcane term for what amounts to development under clear, specific, and actionable guidance to ensure that the model has the grounds to make fair and accurate predictions. For instance, a model may excel in predicting loan defaults; however, if it also accurately predicts non-defaults, it may indicate an issue of unfairness that necessitates a correction.

7.5 REAL-WORLD CASE STUDIES

7.5.1 Case Study 1: Hiring Algorithms

Recruiting is increasingly using AI, with many systems now offering faster and more efficient ways of screening the many resumes that land in employers' inboxes. Although these systems are often viewed as "objective" and even "better" because they can process more data more quickly than humans can, they are just as biased as the humans who created them. They are doing it on a scale that we cannot comprehend and at a speed that we cannot control.

7.5.1.1 Problem

A famous episode occurred when a massive corporation used an algorithm to favor a male-coded language in the resumes examined.

The outcome of the algorithm also reflected the hiring managers' preference for particular colleges, even when the same types of students attended the schools in question. Finally, the algorithm was set to "punish" women's colleges despite the presence of well-qualified female candidates.

7.5.1.2 Explainability Application

SHAP was used to examine the input variables used by the model and how they were used. It turned out that basic signals were causing the model to predict in a way that we did not want. For example, the model associated certain names with a certain gender (in this case, female). When it saw a female name, it pushed the outcome in a way that assumed that the female was more likely to perform whatever action we were trying to predict than if it had seen a male name. Disturbingly, when we dug into it a bit more, we found that the model really liked this gender name signal and was using it instead of—or in addition to—any more reasonable, job-relevant signals it might have been using.

7.5.1.3 Outcome

Equipped with these clues, the team attempted several corrections to the model. They provided the model with a new, more balanced dataset, which was supposed to give equal weight to sex, race, and school type. They also fed several new inputs and set up "fairness" rules that aimed to stop any group from getting an unjust boost. These steps probably made the model less likely to lean on biased cues.

In practice, the updated model gives a more even score across different kinds of applicants, and it passed external fairness audits. Using tools that explain AI decisions seems to be a step toward building truly "ethical" systems. However, these systems are working with a fundamentally flawed dataset. Table 7.3 shows an overview of a case study on hiring algorithm and explainability.

TABLE 7.3
Overview of a Case Study on Hiring Algorithms and Explainability

Aspect	Description
Domain	Recruitment/human resources
Use Case	AI-based resume screening
Identified Problem	Algorithm-favored male-coded language; penalized resumes from women's colleges
Bias Detected In	Name, gendered terms, type of educational institution
Explainability Techniques Used	SHAP (feature importance), counterfactual analysis
SHAP Insights	Name and gendered terms had high influence on selection outcomes
Counterfactual Insight	Changing gender in resumes led to different decisions
Remediation Actions	Retraining model with debiased data; applying fairness constraints
Outcome	Improved fairness across demographics; passed external fairness audits

TABLE 7.4
Overview of Case Study on Loan Approval Systems and Explainability

Aspect	Description
Domain	Finance/banking
Use Case	AI-based loan approval system
Identified Problem	Applicants from minority neighborhoods faced disproportionately high rejection rates
Bias Detected In	Zip code, last names (linked to ethnicity)
Explainability Techniques Used	LIME (local interpretation), SHAP (feature attribution), fairness audits
LIME and SHAP Insights	High influence of zip code and surnames on model predictions
Fairness Audit Findings	Strong correlation between model outputs and protected attributes like race and ethnicity
Remediation Actions	Removed proxy features (e.g., zip code); retrained model; implemented fairness monitoring
Outcome	Improved demographic parity in loan approvals; continuous fairness checks in place

7.5.2 Case Study 2: Loan Approval Systems

Numerous banks employ AI to judge the creditworthiness of their loan applicants. This crucial step in the loan approval process traditionally requires a human decision maker. If an AI system is operating under the modus of "amplifying," as many critics allege, then it is highly possible that some human supervisors at these banks are also working under a similarly biased "amplifying" mode. In effect, the use of AI has covered some possibly unfair lending practices. Table 7.4 summarizes a case study on loan approval systems and explainability.

In particular, a large bank has put in place a ML system to make decisions on whether to approve loan applications. Over time, external parties began to observe that some aspects of the system—some patterns—seemed to be off. Most notably, the system denied loans to a group of applicants, which seemed to have a bias. When the model's decisions were probed with interpretability tools, such as LIME and SHAP, it became apparent that whatever influenced the system to make these biased decisions was not well understood. Zip codes were almost certainly a proxy for the racial composition of particular neighborhoods in the United States. When it came to last names, some were more likely to receive unfavorable decisions than others, and the differences could not be adequately justified by the actual creditworthiness of the individuals.

Additional fairness audits confirmed these findings by showing strong correlations between the predictions made by the model and the protected attributes that define a person's identity, such as race and ethnicity. The strongest evidence came from examining the model's predictions in relation to where people lived. Being from a certain zip code that has been historically linked to underrepresented communities seemed to result, in a systematic way, in being predicted to have a lower likelihood of paying back a loan.

The development team took corrective actions. First, the team stripped out proxy clues, such as zip codes and last names, by changing feature weights. They elevated the status of variables that were supposed to ensure balance in the model and lowered the status of variables that should not influence predictions. Then, they input the data into the fairness algorithm. Now, the model seems to make fairer (less biased) loan decisions, reducing gaps with respect to age and race and helping to regain clients' trust. However, supporters of the model maintain that accuracy is not sacrificed for fairness. Improvement in a performance metric, we are told, should be seen as a good thing. However, what is good? Is it better to have a model that is more accurate but possibly unfair or a model that is somewhat less accurate but much fairer?

7.5.3 Case Study 3: Healthcare Diagnostics

Hospitals are beginning to use AI to help doctors diagnose skin cancer using photographs. In addition to standard photographs, we worked with AI on close-up shots. The team began investigating the tool's performance when the staff raised concerns about its use in dark-skinned individuals. They feared that if the AI was indeed "missing" dark-skinned moles, then the outcome would amount to a missed cancer diagnosis in a segment of the population that already experiences significant health disparities. To find the underlying cause of the issue, we used an explainability tool called saliency maps, which reveal the areas of the image that the AI is using to make its decision. The saliency maps indicated that the AI did not look at the moles in dark-skinned individuals very much at all.

It was looking at random parts of the person's body that had nothing to do with the key diagnostic feature—in other words, the AI was not actually being effective and its performance was shaky. In addition to fixing the data, the nurses and doctors also received a class on how the AI works and its limits. They learned to read the maps and to combine the computer's advice with their own judgment. This may lead to fairer and safer diagnoses if the system is not perfect and clear. Table 7.5 summarizes case study 3.

Equitable outcomes are the end goal of AI applications. However, achieving this goal requires a simultaneous focus on two objectives: fairness and XAI.

Fairness is an often sought-after goal in algorithmic systems. We want systems to be fair, which essentially means that the systems deliver outcomes that pass some kind of fairness test. We also require explainable algorithms. When it comes to systems that have a significant impact on people's lives (and even those that do not but are nonetheless widely used), practitioners want to understand how a system makes decisions. This is where the two objectives are linked.

To build just and trustworthy AI systems:

- Development pipelines must be integrated with explainability.
- Audits must stretch beyond accuracy and social impact.
- Human oversight and regulatory compliance must be part of the practice.

Evolving AI systems can only be designed to be transparent, fair, and accountable if their designers want the benefits of these systems to be distributed widely and not concentrated in the hands of a few. Otherwise, the benefits will probably flow as they always have, toward the already wealthy and powerful people. Table 7.6 summarizes explainability techniques and their fairness applications.

TABLE 7.5
Tabular Representation: Case Study 3

Aspect	Details
Problem	The AI system showed reduced accuracy in diagnosing skin conditions for patients with darker skin tones.
Explainability Application	• Saliency maps revealed that the model focused on non-relevant areas for darker skin. • Analysis exposed underrepresentation of diverse skin tones in the training dataset.
Outcome	• A more diverse dataset was collected and used to retrain the model. • Clinicians received training to interpret AI insights using explainable interfaces.

TABLE 7.6
Summary of Explainability Techniques and Their Fairness Applications

Technique	Purpose	Fairness Use Case
SHAP/LIME	Local feature attribution	Detect influence of sensitive variables
Counterfactuals	Test fairness through input tweaks	Gender/race bias in hiring or lending
Saliency Maps	Visualize model attention (images/text)	Identify demographic biases in medical imaging
Fairness Audits	Holistic system checks	Regulatory compliance in decision pipelines

7.6 EVALUATION METRICS AND BENCHMARKING OF FAIR-EXPLAINABLE AI SYSTEMS

To ensure that AI is both just and comprehensible, we cannot merely correct responses. More may ride on the accuracy of a model's predictions than we have yet to realize. It is conceivable and, many believe, very likely that an AI with high accuracy could still reinforce social inequities if it is trained on or deployed in a manner that is not fair. Therefore, we want to conduct checks, such as measurement equality. We need to know if AI treats different demographic groups similarly. At the same time, we cannot have checks that are easily sidestepped or gamed. With these measures, AIs must also be comprehensible to humans before they can serve as trustworthy digital assistants.

Recent work seems to push for joint scores that balance fairness, transparency, and predictive power. Benchmarks like the COMPAS criminal-justice set, the Adult income data, and the MIMIC-III health records are often used. However, these sets might not fit every context because cultural rules and laws change what fairness looks like.

Therefore, standardized tests that include human feedback, repeatable checks, and cross-field trials could help make AI more accountable. In conclusion, mixed-method evaluation may be the best approach for responsibly rolling out AI in real life.

EXERCISES

1. How does the opaque nature of GenAI contribute to hidden or unintentional bias?
2. In what ways are fairness and explainability interconnected in AI systems?
3. Can improving explainability help identify and mitigate algorithmic bias? If so, how?
4. What challenges arise when both fairness and explainability are lacking in a GenAI system?
5. How can we quantify the impact of model opacity on the fairness outcomes?
6. What role does explainability play in the fairness audits of GenAI models?
7. What tools and techniques can uncover biased decision-making using explanation methods?
8. How can LIME, SHAP, or attention-based explanations reveal demographic or data-driven biases?
9. What are the limitations of current explainability techniques in fairness auditing?
10. How can the findings from explainability be integrated into corrective model retraining?
11. What are the current legal and regulatory requirements for fairness and explainability in AI?
12. How do global regulatory frameworks, such as the EU AI Act and the GDPR, address bias and explainability in GenAI?

13. Can you provide real-world examples or case studies where a lack of fairness or explainability resulted in harmful outcomes?
14. What lessons can be learned from industry case studies regarding the integration of fairness and explainability?
15. How are organizations responding to regulatory pressures through internal policies and governance models?
16. How do HITL systems enhance transparency and fairness in the GenAI?
17. What role does human oversight play in interpreting, validating, and challenging AI-generated outputs?
18. What are the best practices for designing HITL systems that ensure accountability for model decisions?
19. How can stakeholder input be integrated into the GenAI feedback loops?
20. What are the limitations or risks associated with relying on HITL frameworks?

REFERENCES

Amini, A., Soleimany, A., Schwarting, W., Bhatia, S. N., & Rus, D. (2019). Uncovering and mitigating algorithmic bias through learned latent structure. *Proceedings of the AAAI Conference on Artificial Intelligence (AAAI 2019)*, Vol. 33, 622–627. Association for the Advancement of Artificial Intelligence (AAAI Press). https://doi.org/10.1609/aaai.v33i01.3301622

Barocas, S., Hardt, M., & Narayanan, A. (2019). *Fairness and Machine Learning: Limitations and Opportunities.* http://fairmlbook.org

Binns, R. (2018). Fairness in Machine Learning: Lessons from Political Philosophy. *Proceedings of the 2018 Conference on Fairness, Accountability and Transparency*, 149–159.

Chawla, N. V., et al. (2002). SMOTE: Synthetic Minority Over-Sampling Technique. *Journal of Artificial Intelligence Research*, 16, 321–357.

Doshi-Velez, F., & Kim, B. (2017). Towards a Rigorous Science of Interpretable Machine Learning. *arXiv preprint arXiv:1702.08608.*

Ehsan, U., Liao, Q. V., Muller, M., Riedl, M. O., & Weisz, J. D. (2021). Expanding Explainability: Towards Social Transparency in AI Systems. *Proceedings of the 2021 CHI Conference on Human Factors in Computing Systems*, 1–19.

Hardt, M., Price, E., & Srebro, N. (2016a). Equality of Opportunity in Supervised Learning. *Advances in Neural Information Processing Systems*, 29.

Hardt, M., Price, E., & Srebro, N. (2016b). Equality of Opportunity in Supervised Learning. In D. D. Lee, M. Sugiyama, U. V. Luxburg, I. Guyon, & R. Garnett (Eds.), *Advances in Neural Information Processing Systems (NeurIPS 2016)*, Vol. 29, 3315–3323. Curran Associates, Inc. https://proceedings.neurips.cc/paper/2016/hash/9d2682367c3935defcb1f9e247a97c0d-Abstract.html

Kamiran, F., & Calders, T. (2012). Data Preprocessing Techniques for Classification without Discrimination. *Knowledge and Information Systems*, 33(1), 1–33.

Lundberg, S. M., & Lee, S. I. (2017a). A Unified Approach to Interpreting Model Predictions. *Advances in Neural Information Processing Systems*, 30.

Lundberg, S. M., & Lee, S. I. (2017b). A Unified Approach to Interpreting Model Predictions. In I. Guyon, U. V. Luxburg, S. Bengio, H. Wallach, R. Fergus, S. Vishwanathan, & R. Garnett (Eds.), *Advances in Neural Information Processing Systems 30 (NeurIPS 2017)*, 4765–4774. Curran Associates, Inc.

Madaio, M. A., Stark, L., Wortman Vaughan, J., & Wallach, H. (2022). Assessing the Fairness of AI Systems: AI Practitioners' Processes, Challenges, and Needs for Support. *Proceedings of the 2022 ACM Conference on Fairness, Accountability, and Transparency (FAccT)*, 666–678.

Mehrabi, N., Morstatter, F., Saxena, N., Lerman, K., & Galstyan, A. (2021). A Survey on Bias and Fairness in Machine Learning. *ACM Computing Surveys*, 54(6), 1–35.

Miller, T. (2019). Explanation in Artificial Intelligence: Insights from the Social Sciences. *Artificial Intelligence*, 267, 1–38.

Morley, J., Floridi, L., Kinsey, L., & Elhalal, A. (2021). From What to How: An Initial Review of Publicly Available AI Ethics Tools, Methods and Research to Translate Principles into Practices. *Science and Engineering Ethics*, 27(4), 1–31.

Ribeiro, M. T., Singh, S., & Guestrin, C. (2016). "Why Should I Trust You?": Explaining the Predictions of Any Classifier. *Proceedings of the 22nd ACM SIGKDD International Conference on Knowledge Discovery and Data Mining*, 1135–1144.

Selbst, A. D., Boyd, D., Friedler, S. A., Venkatasubramanian, S., & Vertesi, J. (2019). Fairness and Abstraction in Sociotechnical Systems. *Proceedings of the Conference on Fairness, Accountability, and Transparency (FAT*)*, 59–68.

Wachter, S., Mittelstadt, B., & Russell, C. (2017). Counterfactual Explanations without Opening the Black Box: Automated Decisions and the GDPR. *Harvard Journal of Law & Technology*, 31(2), 841–887.

8 Challenges and Limitations

AI has seemingly transitioned from a niche research focus to the core of today's technological, financial, and social infrastructure. AI is ubiquitous in hospitals, banks, schools, traffic control systems, and even governmental decision-making (Lipton, 2018). Because it has permeated so much of society, we must closely examine the potential problems of AI.

AI may offer brighter and sharper results and may even be the basis for novel, "game-changing" ideas, but it is not without limits (Floridi et al., 2018). First, there is a trade-off between the kinds of fairness we want—something that is easy to explain (or not)—and what simply appears to be "wrong" or "right"—and accuracy, which is very expensive to achieve (Jobin et al., 2019).

Computing costs limit the size of models, which limits the types of partnerships that can be formed to solve big problems (e.g., "fairness" versus "raw" accuracy). These are some of the push-and-pull dynamics that drive current developments in AI (Mitchell et al., 2019). We should also consider the legal, social, and ethical aspects of AI. Progress in these areas requires genuine teamwork, as the AI community is not the only necessary player in this field.

8.1 TRADE-OFFS BETWEEN FAIRNESS, EXPLAINABILITY, AND ACCURACY

AI faces a key issue: reconciling multiple goals. In practice, the real systems we build often find themselves in conflict with the goals of transparency, fairness, and performance. When one dimension is optimized, it often comes at the expense of another. This raises the difficult question of whose optimality should be prioritized.

8.1.1 The Accuracy

The fairness dilemma arises in AI when one strives to provide accuracy being sought. This may be a problem with the data, which may possess prejudices that antedate AI, such as gender bias in hiring or racial bias in sentencing (Mehrabi et al., 2021). For instance, some hiring tools may have been trained on past hiring records that favored men, thus treating male and female candidates differently from each other. These systems could certainly find "successful" profiles when reading resumes, yet were also noted to be scoring prospective female candidates with undesirable diminishing returns (Dastin, 2018). In contrast, some medical AI systems trained primarily or solely on data from Western patients often miss predicting several health issues for the same racial minority populations that may have been underrepresented in the training set (Obermeyer et al., 2019). Figure 8.1 depicts fairness dilemmas.

 DOI: 10.1201/9781003713951-8

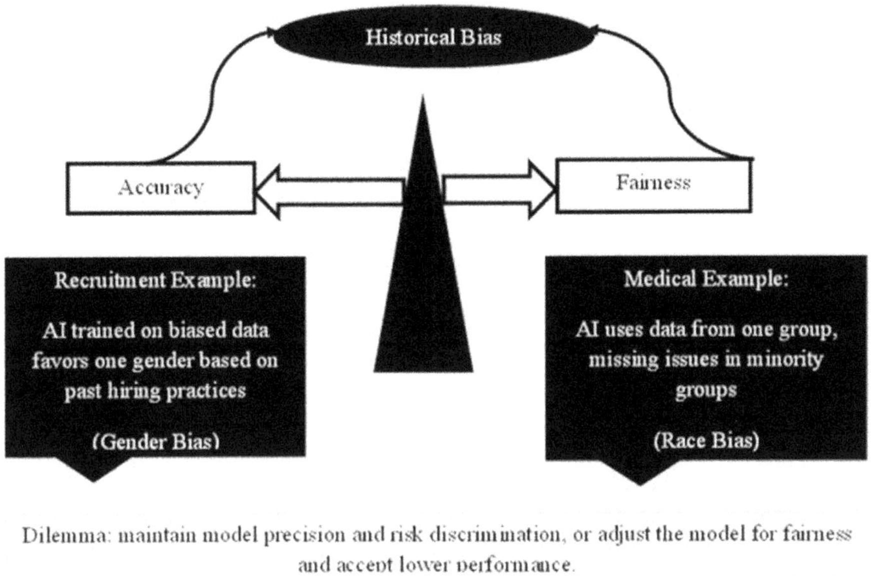

FIGURE 8.1 Fairness Dilemma in AI

8.1.2 Explainability versus Accuracy Model

Prediction accuracy and model interpretability are often seen as conflicting goals, especially regarding the types of models used in ML (Doshi-Velez & Kim, 2017). On the one hand, there are relatively simple and interpretable models, such as decision trees and linear regression. These are not difficult for humans to understand. On the other hand, there are more accurate (and often somewhat more powerful) nonlinear models that generally do not provide a straightforward understanding of how they make predictions (Lipton, 2018). ML models in the latter category include boosted decision trees, support vector machines, and multilayer neural networks (DL) (Wachter et al., 2017). In human-comprehensible terms, these models work in a way that is not easy to describe (or even envision for a human mind) without resorting to a kind of flag-waving for one or another powerful mathematical tool.

8.1.3 Fairness versus Explainability

However, transparency does not automatically result in fair outcomes. A model can provide understandable reasons for a prediction and still make unfair decisions because it is trained on biased data. For example, a decision tree that is used to approve loans could make use of "zip code" as a variable; this seems quite transparent, but it may disadvantage many minority neighborhoods (Barocas et al., 2019). Data from the US Census Bureau show that "Household net worth varies widely by geographic area." However, our adversaries tend to inhabit all parts of the same geography (Edwards & Veale, 2017). This is because the model mentioned at the beginning of this study is quite good. It has a high false-positive rate; however, it also has a decent detection rate.

8.1.4 Multi-Objective Optimization Approaches

To resolve these conflicts, researchers have investigated the concept of multi-objective fairness. They want to strike a balance between fairness and accuracy—between what is easy to explain and what is difficult to explain. They suggested a few different ways to conceptualize this trade-off.

- *Pareto Methods*: Searching for situations in which improving one thing makes another thing worse (Narayanan, 2018).
- *Hybrid Models*: Combining interpretable models with learning models, like networks plus SHAP or LIME (Ribeiro et al., 2016).
- *Fairness-Constrained Learning*: Building algorithms that aim for high accuracy while maintaining limits on how unfair they can be (Zafar et al., 2017).

8.2 COMPUTATIONAL BARRIERS

Significant strides have been made in AI recently, particularly in the areas of ML and DL; however, they carry larger computing loads along with them. These loads are now extending well beyond the pure tech realm and are impacting our environment, finances, and overall accessibility to AI. It is now becoming clear that the sustainability of these colossal new models is an urgent issue that researchers building them—and all of us, really—need to start grappling with today.

8.2.1 Resource-Intensive Training

Building modern AI requires huge computers. Take GPT-3, it has approximately 175 billion knobs, and training it consumed hundreds of petaflop-days. This number may sound absurd, yet it is real (Brown et al., 2020).

Energy use is another issue. Strubell et al. (2019) estimated that one LLM could dump over 626,000 pounds of CO_2 into the air. That is approximately the same amount of carbon that a family of five cars would pollute in a lifetime. Therefore, the climate impact appears to be serious.

Hardware also ties the entire picture together. To teach these models, special chips such as GPUs or TPUs are required. These pieces are expensive and not widely available, which may widen the digital gap (Ahmed & Wahed, 2020).

Some argue that cloud services can share resources, but price tags still favor large firms. In short, only wealthy laboratories can afford the newest AI, which threatens fairness. Policy changes may be helpful in this regard.

8.2.2 Scalability Challenges

Growing datasets may make AI systems increasingly difficult to scale AI systems. One main problem concerns memory and storage; AI requires hardware that can hold terabytes or even petabytes of data. This appears to present a bottleneck for small colleges and other institutions without sufficient resources (Lake et al., 2017). Another issue is the distributed training. While using many GPUs or cloud servers seems necessary for handling the massive data and model sizes, doing so adds complications, latency, and extra communications that may defeat the purpose of parallel processing (Dean et al., 2012). The second is the cost of inference. Running models for tasks that need to happen in real time, such as live translation or figuring out how to keep a self-driving car on the road, seems like it will require money and a low-latency throughput that will likely be a bottleneck for common platforms (Narayanan et al., 2021).

8.2.3 Data Limitations and Quality Constraints

Learning from computers can be made to expand tremendously, but the content we provide restricts their potential. Feeding low-quality data leads to overfitting of such noisy bits. If the training set is skewed, we may amplify the bias. This results in incorrect predictions at the other end of the process, which can waste computing cycles if not careful. Of course, producing all this labeled data with a large enough workforce and sufficient cash to keep that workforce happy is a significant and often-ignored bottleneck in research (Geiger et al., 2021; Mehrabi et al., 2021).

8.2.4 Environmental Sustainability Concerns

AI's potential climate impact seems quite concerning because large models require a lot of power, often drawn from dirty energy sources. For example, consider the BERT model.

- Training BERT is roughly equivalent to the energy consumed by a "whole aircraft traveling cross-country," as the researchers wrote in a 2019 paper (Bessen, 2019).
- Meanwhile, on the climate front, a 2018 report pointed out that "data centers that run AI already consume a large and growing amount of global electricity."

Some observers weigh the costs and benefits and decide that the balance still tips in favor of the planet. However, this energy consumption does not feel like "renewable progress."

8.2.5 Accessibility and Global Divide

Geopolitical gaps in AI access may pose another hurdle. The United States and China keep pushing huge models, yet many newer countries seem stuck because of the following challenges:

- No high-performance computing gear
- Cloud services too expensive for their budgets
- Companies depend on foreign AI tools (Crawford, 2021).

While other nations may endeavor to equalize, they are still far behind in funding. The scene may tempt us to imagine some team effort, yet the competition is fierce. Students in these locales often miss out on what one might call the apotheosis of American Science, Technology, Engineering, and Mathematics (STEM) education, deepening our regional skill divide. This imbalance might lead to a tiny AI lite in America, which could widen the worldwide tech divide (Zuboff, 2019).

8.2.6 Potential Mitigation Strategies

Computing problems that plague the scientific community and professional sectors are being targeted for solutions to these problems. Some of the most hopeful appear to be:

- *Model Compression*: Han et al. (2015) posit that schemes such as pruning, quantization, and knowledge distillation can make a model work less hard for the same accuracy.
- *Efficient Architectures*: Tan and Le (2019) describe networks such as MobileNets and EfficientNet as potentially taking a hit in performance but using far fewer resources than their contemporaries.

Schwartz et al. (2020) warn against equating model efficiency with getting high accuracy. They allow that a more formalized way to judge a model might include how well it works given the kinds of resources available and its "carbon footprint."

8.3 LEGAL BARRIERS

Increasingly, we find AI acting in our world—hospitals, banks, hiring offices, city halls. This may result in legal complications. AI is different from the programs we know and love (or loathe). It performs tasks without our direct input and in ways that can only be understood by the AI itself (and perhaps not even by it all the time). When things go wrong—and they will—we might be hard-pressed to say who is responsible. No worldwide laws exist, which likely hurts the trust of everyday people.

8.3.1 Regulatory Fragmentation and Inconsistency

One basic roadblock is that the world does not have a single rule set for AI. Different places act in their own unique ways. The EU has proposed the Artificial Intelligence Act of 2021, which categorizes systems into three risk groups: unacceptable, high, and low (Floridi, 2021).

- The United States appears to fragment the field; rules live in each sector instead of a national law (Calo, 2017).
- China pushes for strong state control, tying AI to its big national plans (Ding, 2018).

Therefore, firms that sell worldwide face a clash of rules that keep changing, which may hurt their planning for many markets worldwide today (Butcher & Beridze, 2019).

8.3.2 Liability and Accountability

Perhaps we do not know how to assign legal blame when AI hurts people. Some say that the rulebook may not fit technology. For example:

- Could a programmer be held responsible if a self-driving car crashes?
- If an algorithm makes a bad trade, who carries the loss: its creators, the firm using it, or the regulator?

The traditional tort system assigns responsibility to individuals for their harmful actions; however, the increasing use of AI pushes decision-making to a higher level, making it unclear who is at fault (Gless et al., 2016). When a smart machine causes an injury, who is held accountable? Until tort law is reformed to address such questions, it has an unfinished feel, as if it were designed only for the past and not for the present or future.

8.3.3 Privacy and Data Protection

AI requires a lot of data, often of a personal nature, which is a genuine cause for concern over privacy. Following the lead of the EU's GDPR, companies cannot use our data without our consent, and even then, they must tell us how and for what purpose they are using it—and this applies to any automated decisions made by AI. If the cloud is our computing future, then to enforce the law in AI's case, we must have a way to keep all of this cross-border data under the various jurisdictions of the different countries it might pass through (Buiten, 2019). Otherwise, do you really think the private sector will behave in a paternalistic way for our privacy?

8.3.4 Intellectual Property Rights

Traditional intellectual property (IP) concepts become entangled in the web of complications woven by AI. This occurs in several ways. Can works (texts, images, music) produced by AI be protected by copyright? If so, who holds the copyright? Is it the developer, user, or AI program itself? Then we have patents. Should the name of an AI program be included as an inventor on a patent? What does it mean for such an entity to be recognized as an inventor, particularly when the program is working under the direction of a large corporation? Finally, we have the use of data to train AI. Copyright law is implicated here, with the potential use of data for "fair use" purposes and in the much thornier area of data licenses (Burrell, 2016).

8.3.5 Algorithmic Bias and Anti-Discrimination Laws

AI program bias may breach discrimination laws, particularly those related to employment, credit, and policing. For instance, consider the following three contexts:

Appointments: Amazon's automated hiring program was found to discriminate against female candidates (Dastin, 2018). It is employed by the American judiciary system but has faced backlash for its racial biases in estimating recidivism rates (Angwin et al., 2016).

Financial Services: AI-based credit scoring could reinforce structural bias if models rely on historically discriminatory data. Although the Civil Rights Act and EU laws of nondiscrimination exist, enforcing them against black-box AI models is difficult (Barocas & Selbst, 2016).

8.3.6 Lack of Explainability in Legal Contexts

The law often demands that certain things be understood and clarified. For example, the GDPR requires organizations to provide an explanation when they make decisions about people using only automated processes without human intervention. In legal matters, it is essential that all parties comprehend the reasoning behind a decision that directly impacts them.

However, many DL models, which are most likely to be used for making important automated decisions about people, are functionally opaque. They do not have a human-understandable way of showing a plausible reason for their decision (Goodman & Flaxman, 2017). And that, of course, is a prerequisite for any decision that you want to uphold as legal.

8.3.7 Toward AI-Specific Legal Frameworks

Advocates in research and government call for a legal environment that is better suited to AI. They want a future in which innovation in AI can proceed but under conditions that assure societal stakeholders that AI will not endanger them or their jobs. Several ideas for achieving harmony have recently emerged. For instance, a group at the Berkman Klein Center for Internet and Society at Harvard examined whether the current law is sufficient to cover the most important domains of life where autonomous agents might cause harm (Reisman et al., 2018). They proposed creating a new, preemptive review mechanism they call "algorithmic impact assessments."

8.4 SOCIAL BARRIERS

AI does not operate in a vacuum; it is used by people, and, at the most fundamental level, people cannot exist without trust and ethics. While it seems inevitable that AI will bring about enormous changes, there are also many social blocks that stop those changes either from being realized or from being realized too quickly. They include general doubt of AI, internet access that is as uneven as access to a good education, fear of lost jobs, and a variety of cultural differences that together constitute a very untrusting society (Jones, 2018).

8.4.1 Public Trust and Acceptance

The factors driving AI adoption include trust, which many people feel is lacking. A high level of trust is required for AI to be adopted and used in high-stakes domains such as healthcare, criminal justice, and finance. Public fear and concern over automation by AI could inhibit trust from developing (Brynjolfsson & McAfee, 2017). Accidents involving the most talked-about autonomous systems, such as the Uber vehicle that killed a pedestrian in 2018 (Goodall, 2014), have bred public skepticism toward using AI for anything beyond predetermined sets of instructions.

Even when AI makes decisions that could be expected to drive public trust upward, such as the decision-making that occurs in a self-driving car, the "black-box" nature of many current AI models tends to foster opacity in a way that pushes trust back down. Improving public trust in AI could enhance the public's embrace of future AI applications.

8.4.2 Workforce Displacement and Economic Inequality

The automation of work by AI has sparked fears about the potential for mass unemployment and income inequality. Scholars have estimated that as much as 47% of US work might be displaced by robotics and AI in the coming decades (Frey & Osborne, 2017). However, even if the obsolescence of a specific job does not lead to labor market polarization, the skills gap can. Economies in developed countries lack the retraining programs necessary to prepare displaced workers for AI-augmented jobs in the future. We are on track to see the same widening economic divide that has appeared since the advent of personal computing.

8.4.3 Digital Divide and Inequitable Access

The uneven adoption of AI across societies reflects the existing digital inequalities. On a geographic front, the latest AI infrastructure is concentrated in the Global North—North America and Europe, and some East Asian nations—leaving much of the developing world behind (UNESCO, 2021). When examined from a socioeconomic perspective, many marginalized communities in the United States and worldwide still lack access to AI-driven services in healthcare, education, and finance (Eubanks, 2018). Add to this mix the representation problem: Minority groups in the United States and around the world are underrepresented in the kinds of training datasets that form the basis of AI (Noble, 2018).

8.4.4 Ethical Concerns and Social Acceptance

AI raises forthright ethical concerns that stand in the way of public acceptance.

- *Surveillance and Autonomy*: The use of AI for facial recognition has spurred intense debate over privacy and governmental monitoring (Harwell, 2020).
- *Moral Responsibility*: Giving AI the power to make pivotal life-and-death choices for humans, as it does with military drones and in some medical decision-making, is hard to swallow ethically (Sharkey, 2019).
- *Cultural Perspectives*: Human supervision is more valued than machine autonomy in some cultures, which also affects perspectives on acceptance (Hagerty & Rubinov, 2019).

To overcome these ethical roadblocks, we must place more human value on the architecture and governance of AI.

8.4.5 Cultural and Regional Differences

These cultural contexts strongly shape societies' ways of seeing and using AI. Most Western cultures focus on individual rights, process clarity, and personal responsibility. "Yes, we will adopt AI," say these cultures, "but first, we need to discuss and debate individual and collective rights to ensure regulatory clarity and oversight." In contrast, Japan and South Korea have not engaged in such debates. They live with AI-promoting robots and automata in their homes and on the streets. For these nations, AI adoption is as natural as that of a humanoid robot (Robertson, 2017).

8.4.6 Media Influence and Public Discourse

Public opinion on AI is largely formed by mass media. The media's coverage of AI could even be seen as one of the factors amplifying or dampening concerns over the technology (Cave et al., 2019).

Newspapers tend to alternate between portraying AI as a heroic technology that will make our future health (such as curing diseases) and as a dangerous force that could create massive unemployment or take over the world (West, 2018). AI has been framed as both a tool and a threat. Fair coverage of technology in newspapers is of paramount importance for ensuring that substantive public discussions on AI take place.

8.4.7 Strategies for Overcoming Social Barriers

Several paths can be taken to resolve the social resistance many people have toward AI.

The first is participatory design. This method was developed by the Scandinavian school of human–computer interaction, which focuses on engaging users not just in the testing phase, as is traditional with usability studies but also in the design and implementation phases of a project (Lee et al., 2019). The understanding gained from working closely with end users in this way is invaluable and leads to a much greater sense of trust and ownership.

The second strategy is clarity and understandability. In a world with so many complex systems, if you are going to design one that people have to live with, it should be as straightforward as possible. In my opinion, this is not just about making AI systems explainable. At its core, I think what we are getting at is human acceptability.

8.5 STANDARDIZATION GAPS IN EVALUATION

AI tools continue to appear in important, sensitive areas, even though no worldwide rules exist. Unlike older software that can be measured against clear outcomes, many AI systems act like black boxes, making their choices difficult to see. This may raise questions about fairness, strength, and the ability to explain (Kairouz et al., 2021). Therefore, trust between users and manufacturers can decrease. Because standards are missing, careful and safe AI use is often postponed.

8.5.1 Fragmented Benchmarks and Metrics

Currently, AI assessments rely on domain-specific metrics that are often inconsistent or outdated.

- *Benchmark Saturation*: The models perform nearly at par with or better than humans on widely used datasets (e.g., ImageNet, Modified National Institute of Standards and Technology-MNIST database); however, these datasets fail to encapsulate the true variety of real-world scenarios (Recht et al., 2019).
- *Fairness and Diversity Measures*: Some fairness metrics conflict with one another, such as demographic parity and equal opportunity, which models cannot satisfy all at once (Kleinberg et al., 2017).
- *Evaluation Metrics*: There are some overall metrics that exist to quantify interpretability, but most are dependent on the specific problem at hand and are rather subjective (Doshi-Velez & Kim, 2017).

This scattershot approach prevents direct comparisons between different systems and calls their reproducibility into question.

8.5.2 Lack of Fairness Standardization

Ambiguity in implementation arises from the absence of a uniform definition of fairness. Is it fair for people who operate in different social domains to be treated differently? Or is fairness about equal treatment, as claimed by US anti-discrimination law? Even if we could agree, in principle, on what is fair, we would still need to contend with trade-offs: If we measure fairness with respect to certain dimensions, we can achieve high scores; however, when we do this, we diminish accuracy in the process. Fairness demands that a high level of performance be achieved (Corbett-Davies et al., 2017). Because jurisdictions differ in their understanding of fairness, organizations operating across jurisdictions cannot use a common set of fairness criteria.

8.5.3 Explainability and Interpretability Challenges

The missing piece in responsible AI is a standardized way to assess the explainability of an AI model (Slack et al., 2020). Currently, we use a few model-agnostic tools that have some advantages and disadvantages. LIME and SHAP: Both offer an interesting way to think about local versus global models (Miller, 2019). I believe that each has its strengths as well as some weaknesses. These tools can be utilized to evaluate the extent to which an AI model can be explained (Wachter et al., 2017). However, what occurs when a model exhibits deficiencies, rendering it inexplicable in a coherent manner? Is the distinction between local and global models beneficial in this context? Can you identify any additional tools?

8.5.4 Reproducibility Crisis in AI Research

A serious obstacle is that the work cannot be duplicated (or is poorly duplicated). There are few (if any) source codes, datasets, or hyperparameter setting releases accompanying AI papers, which makes them extremely opaque (Hutson, 2018). Even the best benchmark datasets are poor proxies for the real world, in part because they are "straw men" that must be defeated by the systems trained on them, and in part because they must serve as surrogates for a baser reality (which is itself quite variable over both space and time) (Quiñonero-Candela et al., 2009). Finally, even if one had access to all the real-world data needed to build a truly robust and generalizable model, one would still be unable to duplicate the results of AI without access to a supercomputer (which very few people have) (Ahmed & Wahed, 2020).

8.5.5 Absence of Cross-Domain Evaluation Frameworks

Contemporary AI systems have already been implemented in multi-domain scenarios such as medicine, law, and finance. The reason for this is clear: these disciplines rely heavily on data, which is the lifeblood of any computational intelligence (Kuner et al., 2020). Yet when it comes to discipline-hopping AI—and AI applied in more than one scenario, medicine for example, but also law—its assessment remains the same: largely one-off and essentially place-based. Moreover, such assessments are being conducted even as AI is being engineered for cross-disciplinary utility.

Health AI: Serious AI researchers and engineers value sensitivity and specificity—our own Dr. Topol ranks them as the top two characteristics of good health AI—for measuring performance. Yet, are they valuing these characteristics for the right reasons? Are they valuing them, in other words, simply because those are the appropriate metrics for assessing health AI or because for a variety of undisclosed reasons (ambitions untethered by any binding ethical guidelines or principles), the researchers and engineers in question happen to see them as the best or most scientific way to assess the performance of an AI in a health scenario? Limitations of AI assessment paradigms are shown in Table 8.1.

TABLE 8.1
Limitations of Current Domain-Specific AI Assessment Paradigms

Domain	What Is Commonly Measured	What Is Commonly Neglected	Key References
Health AI	Sensitivity, specificity, diagnostic accuracy	Fairness, interpretability, bias impact, patient-centric outcomes	Topol (2019)
Finance AI	Predictive validity, risk estimation performance	Ethical considerations, transparency, explainability	Morley et al. (2019)
Legal AI	Predictive validity of case outcomes	Social, cultural, and systemic effects; fairness; accountability	Završnik (2020)

8.5.6 Emerging Standardization Efforts

Numerous attempts have been made to address the standardization deficit, but these attempts have also been fragmented. ISO/IEC JTC 1/SC 42 focuses on establishing international standards for AI (Gebru et al., 2018; Mitchell et al., 2019). However, on the mainly voluntary front, the NIST AI Risk Management Framework of 2023 takes a top-down, federal approach to responsible AI that is positive in intent but vague and heavily couched in qualifiers about what not to do.

The IEEE 7000 Series has an even lighter touch, with virtually no enforcement mechanisms but good talks about ethics and not being vague and fluffy that people cannot figure out what it is saying (IEEE, 2020). The closer-in strategies, Model Datasheets, and Model Cards are also highly recommended but remain voluntary.

8.5.7 Toward Unified Evaluation Standards

To address these deficiencies, planners and researchers have suggested a number of remedial actions.

- *Use Multidimensional Metrics*: Do not just evaluate AI on how correct it is; evaluate it on its fairness, explainability, robustness, and sustainability. These are the dimensions on which AI can fail and society cannot afford to fail (Levendowski, 2018).
- *Use Context-Specific Standards*: Regions and the contexts in which AI systems are employed differ. Respecting this variety while adhering to some authority of central ethical principles could lead to better AI standards (Samuelson, 2019).
- *Conduct Independent Audits*: Who watches the watchers? If third parties watched the algorithmic systems being employed in society, it would almost certainly lead to better-behaved systems and better outcomes for society (Raji et al., 2020).
- *Act Globally and Synchronously*: To the extent possible, AI standards should be global and synchronized. Otherwise, we risk a situation in which AI systems are not trustworthy or reproducible.

8.6 CONCLUSION AND FUTURE OUTLOOK

Building AI that is fair, clear, and accountable is not a simple task; you run into tech glitches, legal hoops, social pushback, and institutional quirks. This part of the book shows how achieving fairness, clear explanations, and exact results means making some tough choices that most people will not like; then the technology, the laws, and the way people think all get in the way of actually using it, and because there is no single standard, it hurts the ability to repeat, compare, or trust the work. Fairness and accuracy tug each other, so is there perfect AI equity? This is not feasible; therefore, we must settle for case-by-case compromises (Shneiderman, 2020).

In 2017, things changed rapidly; everything seemed to turn upside down, and I could hardly keep up. Computer challenges also make things tougher because the tools for explainability cannot keep up, and resources are uneven across AI fields; therefore, everything gets harder (Ahmed & Wahed, 2020). Therefore, in law and everyday life, vague rules, cultural fights over what is fair, and a public that still doubts AI prevent the technology from being used in places such as hospital triage or storm alerts (Veale & Borgesius, 2021). Standardization issues? It seems to be everywhere. No global standards exist; therefore, each organization is left guessing and cannot line up AI models to evaluate them. Although the NIST AI Risk Management Framework, IEEE 7000 series standards, and Datasheets for Datasets are new efforts, they are fragmented, optional, and rarely followed in a uniform way. (In *2018,* Mitchell et al. reported their findings. The article appeared in 2019.) Therefore, solving these problems in the near future will require collaboration between governments, universities, businesses, and civil groups. Integrative assessment metrics: it is not only plain accuracy anymore but also fairness, interpretability, robustness, and sustainability bundled together (Vinuesa et al., 2020).

- Therefore, we try using worldwide or local rules that look the same, but they bend to fit the tribe's own laws and customs. Human-centric design? It helps explain stuff and add fair tweaks for all kinds of folks—policymakers, affected neighborhoods, and everyday people.
- *Algorithmic Auditing*: It shows that autonomous external audits can be used to test fairness and responsibility claims (Raji et al.). The 2020 findings show a huge shift, so we ought to rethink our approach.
- Sustainable AI research aims to spread computing power to more people; therefore, it tries to stop tight resource limits from deepening AI inequality (Ahmed & Wahed, 2020).

These limits are not impossible barriers; therefore, design rules are needed to shape responsible AI. Aware of the limits, scientists and specialists might try solutions that hit the mark: practical tech, moral duty, cultural fit. Therefore, the thinking we do now fuels the next chapter: what could AI become, and who should watch over it? After noting the problems, this study looks at hybrid models, symbolic thinking, fairness-focused designs, and governing rules to try and line up current limits with hopes for a fair, trustworthy AI.

REFERENCES

Ahmed, N., & Wahed, M. (2020). The de-democratization of AI: Deep learning and the compute divide in artificial intelligence research. *arXiv preprint arXiv:2010.15581.*

Angwin, J., Larson, J., Mattu, S., & Kirchner, L. (2016). *Machine bias: There's software used across the country to predict future criminals.* ProPublica. https://www.propublica.org/article/machine-bias-risk-assessments-in-criminal-sentencing

Barocas, S., Hardt, M., & Narayanan, A. (2019). *Fairness and machine learning.* fairmlbook.org.

Barocas, S., & Selbst, A. (2016). Big data's disparate impact. *California Law Review, 104*(3), 671–732.

Bessen, J. (2019). AI and jobs: The role of demand. *NBER Working Paper No. 24235.*

Brown, T., Mann, B., Ryder, N., et al. (2020). Language models are few-shot learners. In H. Larochelle, M. Ranzato, R. Hadsell, M. F. Balcan, & H. Lin (Eds.), *Advances in neural information processing systems 33 (NeurIPS 2020)* (pp. 1877–1901). Curran Associates, Inc. https://proceedings.neurips.cc/paper/2020/hash/1457c0d6bfcb4967418bfb8ac142f64a-Abstract.html

Brynjolfsson, E., & McAfee, A. (2017). *Machine, platform, crowd: Harnessing our digital future.* W. W. Norton & Company.

Buiten, M. C. (2019). Towards intelligent regulation of artificial intelligence. *European Journal of Risk Regulation, 10*(1), 41–59.

Burrell, J. (2016). How the machine "thinks": Understanding opacity in machine learning algorithms. *Big Data & Society, 3*(1).

Butcher, J., & Beridze, I. (2019). What is the state of artificial intelligence governance globally? *The RUSI Journal, 164*(5–6), 88–96.

Calo, R. (2017). Artificial intelligence policy: A primer and roadmap. *U.C. Davis Law Review, 51*, 399–435.

Cave, S., Coughlan, K., & Dihal, K. (2019). "Scary robots": Examining public responses to AI. *Proceedings of AIES.*

Corbett-Davies, S., Pierson, E., Feller, A., et al. (2017). Algorithmic decision making and the cost of fairness. *Proceedings of the 23rd ACM SIGKDD International Conference on Knowledge Discovery and Data Mining (KDD 2017)* (pp. 797–806). Association for Computing Machinery (ACM). https://doi.org/10.1145/3097983.3098095

Crawford, K. (2021). *Atlas of AI: Power, politics, and the planetary costs of artificial intelligence.* Yale University Press.

Dastin, J. (2018). *Amazon scraps secret AI recruiting tool that showed bias against women.* Reuters. https://www.reuters.com/article/us-amazon-com-jobs-automation-insight-idUSKCN1MK08G

Dean, J., Corrado, G., Monga, R., et al. (2012). Large scale distributed deep networks. In F. Pereira, C. J. C. Burges, L. Bottou, & K. Q. Weinberger (Eds.), *Advances in neural information processing systems 25 (NIPS 2012)* (pp. 1223–1231). Curran Associates, Inc. https://proceedings.neurips.cc/paper/2012/hash/6aca97005c68f1206823815f66102863-Abstract.html

Devlin, J., Chang, M.-W., Lee, K., & Toutanova, K. (2019). BERT: Pre-training of deep bidirectional transformers for language understanding. *Proceedings of the 2019 Conference of the North American Chapter of the Association for Computational Linguistics: Human Language Technologies (NAACL-HLT 2019)* (pp. 4171–4186). Association for Computational Linguistics. https://doi.org/10.18653/v1/N19-1423

Ding, J. (2018). *Deciphering China's AI dream.* Future of Humanity Institute, University of Oxford.

Doshi-Velez, F., & Kim, B. (2017). Towards a rigorous science of interpretable machine learning. *arXiv preprint arXiv:1702.08608.*

Edwards, L., & Veale, M. (2017). Slave to the algorithm? Why a right to an explanation is probably not the remedy you are looking for. *Duke Law & Technology Review, 16*(1), 18–84.

Eubanks, V. (2018). *Automating inequality: How high-tech tools profile, police, and punish the poor.* St. Martin's Press.

Floridi, L. (2021). The European legislation on AI: A brief analysis of its philosophical approach. *Philosophy & Technology, 34*, 215–222.

Floridi, L., Cowls, J., Beltrametti, M., et al. (2018). AI4People: An ethical framework for a good AI society. *Minds and Machines, 28*(4), 689–707.

Frey, C. B., & Osborne, M. A. (2017). The future of employment: How susceptible are jobs to computerization? *Technological Forecasting and Social Change, 114*, 254–280.

Gebru, T., Morgenstern, J., Vecchione, B., et al. (2018). Datasheets for datasets. *arXiv:1803.09010.*

Geiger, R. S., Yu, K., Yang, Y., et al. (2021). Garbage in, garbage out? Do machine learning application papers in social computing report where human-labeled training data comes from? *Proceedings of CSCW.*

Gless, S., Silverman, E., & Weigend, T. (2016). If robots cause harm, who is to blame? *New Criminal Law Review*, *19*(3), 412–436.

Goodall, N. (2014). Machine ethics and automated vehicles. *Road Vehicle Automation*, *93*, 93–102.

Goodman, B., & Flaxman, S. (2017). European Union regulations on algorithmic decision-making and a "right to explanation." *AI Magazine*, *38*(3), 50–57.

Hagerty, A., & Rubinov, I. (2019). Global AI ethics: A review of the social impacts and ethical implications of AI. *Technology in Society*, *63*, 101258.

Han, S., Mao, H., & Dally, W. J. (2015). Deep compression: Compressing deep neural networks with pruning, trained quantization and Huffman coding. *arXiv:1510.00149.*

Harwell, D. (2020). Facial recognition's threat to privacy. *The Washington Post.*

Hutson, M. (2018). Artificial intelligence faces reproducibility crisis. *Science*, *359*(6377), 725–726.

IEEE. (2020). *IEEE 7000 series standards for AI ethics*. IEEE Standards Association.

Jobin, A., Ienca, M., & Vayena, E. (2019). The global landscape of AI ethics guidelines. *Nature Machine Intelligence*, *1*(9), 389–399.

Jones, N. (2018). How to stop data centres from gobbling up the world's electricity. *Nature*, *561*(7722), 163–166.

Kairouz, P., McMahan, H. B., Avent, B., et al. (2021). Advances and open problems in federated learning. *Foundations and Trends in Machine Learning*, *14*(1–2), 1–210.

Kleinberg, J., Mullainathan, S., & Raghavan, S. (2017). Inherent trade-offs in the fair determination of risk scores. *Proceedings of ITCS.*

Kuner, C., Svantesson, D. J., Bygrave, L. A., et al. (2020). The global privacy assembly: Strengthening international cooperation in data protection. *International Data Privacy Law*, *10*(3), 221–234.

Lake, B., Ullman, T., Tenenbaum, J., & Gershman, S. (2017). Building machines that learn and think like people. *Behavioral and Brain Sciences*, *40*, e253.

Lee, M. K., Jain, A., Cha, H. J., et al. (2019). Participatory design for AI: Designing human-centered AI systems. *CHI Conference Proceedings.*

Levendowski, A. (2018). How copyright law can fix artificial intelligence's implicit bias problem. *Washington Law Review*, *93*, 579–630.

Lipton, Z. C. (2018). The mythos of model interpretability. *Communications of the ACM*, *61*(10), 36–43.

Mehrabi, N., Morstatter, F., Saxena, N., et al. (2021). A survey on bias and fairness in machine learning. *ACM Computing Surveys*, *54*(6), 1–35.

Miller, T. (2019). Explanation in artificial intelligence: Insights from the social sciences. *Artificial Intelligence*, *267*, 1–38.

Mitchell, M., Wu, S., Zaldivar, A., et al. (2019). Model cards for model reporting. *Proceedings of the Conference on Fairness, Accountability, and Transparency* (pp. 220–229). Association for Computing Machinery (ACM). https://doi.org/10.1145/3287560.3287596

Morley, J., Floridi, L., Kinsey, L., & Elhalal, A. (2019). From what to how: An initial review of publicly available AI ethics tools, methods and research to translate principles into practices. *Ethics of Information Technology*, *22*(4), 1–20.

Narayanan, A. (2018). Translation tutorial: 21 fairness definitions and their politics. *FAT Conference.*

Narayanan, D., Phanishayee, A., Shi, S., et al. (2021). Efficient large-scale language model training on GPU clusters using Megatron-LM. *Proceedings of SC '21.*

Noble, S. U. (2018). *Algorithms of oppression: How search engines reinforce racism.* NYU Press.

Obermeyer, Z., Powers, B., Vogeli, C., & Mullainathan, S. (2019). Dissecting racial bias in an algorithm used to manage the health of populations. *Science, 366*(6464), 447–453.

Quiñonero-Candela, J., Sugiyama, M., Schwaighofer, A., & Lawrence, N. (2009). *Dataset shift in machine learning.* MIT Press.

Raji, I. D., Smart, A., White, R. N., et al. (2020). Closing the AI accountability gap: Defining an end-to-end framework for internal algorithmic auditing. *Proceedings of the 2020 ACM Conference on Fairness, Accountability, and Transparency* (pp. 33–44). Association for Computing Machinery (ACM). https://doi.org/10.1145/3351095.3372873

Recht, B., Roelofs, R., Schmidt, L., & Shankar, V. (2019). Do ImageNet classifiers generalize to ImageNet? *Proceedings of the 36th International Conference on Machine Learning (ICML)* (pp. 5389–5400). ML Research Press (via PMLR).

Reisman, D., Schultz, J., Crawford, K., & Whittaker, M. (2018). *Algorithmic impact assessments: A practical framework for public agency accountability.* AI Now Institute.

Ribeiro, M. T., Singh, S., & Guestrin, C. (2016). "Why should I trust you?" Explaining the predictions of any classifier. *Proceedings of KDD.*

Robertson, J. (2017). *Robo sapiens japanicus: Robots, gender, family, and the Japanese nation.* University of California Press.

Samuelson, P. (2019). Should AI-generated inventions be patentable? *Berkeley Technology Law Journal, 34*(2), 385–430.

Schwartz, R., Dodge, J., Smith, N. A., & Etzioni, O. (2020). Green AI. *Communications of the ACM, 63*(12), 54–63.

Sharkey, N. (2019). Autonomous weapons and human responsibility. *International Review of the Red Cross, 94*(886), 701–713.

Shneiderman, B. (2020). Human-centered AI. *ACM Interactions, 27*(4), 76–81.

Slack, D., Hilgard, A., Jia, E., et al. (2020). Fooling LIME and SHAP: Adversarial attacks on post hoc explanation methods. *Proceedings of the AAAI/ACM Conference on AI, Ethics, and Society* (pp. 180–186). Association for Computing Machinery (ACM). https://doi.org/10.1145/3375627.3375830

Strubell, E., Ganesh, A., & McCallum, A. (2019). Energy and policy considerations for deep learning in NLP. *ACL Conference Proceedings.*

Tan, M., & Le, Q. (2019). EfficientNet: Rethinking model scaling for convolutional neural networks. In H. Larochelle, M. Ranzato, R. Hadsell, M. F. Balcan, & H. Lin (Eds.), *Proceedings of the 36th International Conference on Machine Learning (ICML 2019)* (Vol. 97, pp. 6105–6114). PMLR. http://proceedings.mlr.press/v97/tan19a.html

Topol, E. (2019). High-performance medicine: The convergence of human and artificial intelligence. *Nature Medicine, 25*(1), 44–56.

UNESCO. (2021). *Recommendation on the ethics of Artificial Intelligence.* UNESCO.

Veale, M., & Borgesius, F. Z. (2021). Demystifying the draft EU Artificial Intelligence act. *Computer Law Review International, 22*(4), 97–112.

Vinuesa, R., Azizpour, H., Leite, I., et al. (2020). The role of artificial intelligence in achieving the sustainable development goals. *Nature Communications, 11*, 233.

Wachter, S., Mittelstadt, B., & Floridi, L. (2017). Why a right to explanation of automated decision-making does not exist in the GDPR. *International Data Privacy Law, 7*(2), 76–99.

West, D. M. (2018). *The future of work: Robots, AI, and automation.* Brookings Institution Press.

Zafar, M. B., Valera, I., Rodriguez, M. G., & Gummadi, K. P. (2017). Fairness beyond disparate treatment & disparate impact: Learning classification without disparate mistreatment. *Proceedings of WWW.*

Završnik, A. (2020). Criminal justice, artificial intelligence systems, and human rights. *ERA Forum, 21*(4), 567–583.

Zuboff, S. (2019). *The age of surveillance capitalism: The fight for a human future at the new frontier of power.* PublicAffairs.

9 Future Directions and Responsible AI

AI, more specifically GenAI, has seen some important advancements recently. However, AI still has massive and significant interpretive problems (heaps of it interpretive in nature; as the saying goes, "who interprets, governs") that raise interpretive, fairness, governance, and alignment issues. This is for very good reasons. We are interested in all these issues because they are matters of "responsibility," "justice," and "governance." Should we expect something to be responsible if we do not understand how it works? Should we simply trust that it will be fair? Hands up for rhetorical questions! AI has seen understandable governance problems over the last several years and interpretable fairness issues since its inception. Despite significant progress in significant parts of AI and the availability of several communication platforms, we have not yet seen much public engagement or discussion on how to make governance issues understandable to the people responsible for upholding justice and governing society. AI will only complicate this interpretive problem.

9.1 RESEARCH OPPORTUNITIES: HYBRID MODELS AND SYMBOLIC REASONING

Combining statistical ML and symbolic reasoning is a substantial step forward in AI progress. DL systems are excellent at recognizing patterns and not much else. They should not be confused with the type of "intelligence" that humans possess. In fact, when it comes to things like knowledge representation, logical inference, and the kind of sound and foggy reasoning that humans are so good (or not so good) at, symbolic AI is much better than DL. However, what if these two could work together? They may produce a better form of "intelligence" for AI systems. And researchers at Google and elsewhere have been attempting this kind of unification for some time in what is called "hybrid AI."

9.1.1 The Foundations of Hybrid AI

Hybrid AI—also known as neuro-symbolic AI—brings together two ways of doing AI: neural networks and symbolic reasoning. Neural networks are adept at handling complicated, high-dimensional data, such as images, language, and sound. However, if you want to perform real reasoning and decision-making and want your system to be more like a human brain, you must add symbolic mechanisms to the mix.

In a typical hybrid architecture, there are layers of stuff. The lower layers are composed of deep neural networks that perform the heavy lifting necessary to make category judgments and perform feature extraction, which are difficult for humans to figure out (Barocas et al., 2019). The large architectures that perform this type of

DOI: 10.1201/9781003713951-9

work nowadays—such as the one behind this text—have millions or even billions of parameters. These networks typically do not know or care about the kinds of distinctions that are vital for humans to make in order to reason effectively.

9.1.2 Notable Examples and Industrial Adoption

A prime example is IBM's Neuro-Symbolic Concept Learner (NS-CL), which integrates CNNs with logical inference engines to perform visual question-answering tasks. This model decomposes an image into symbolic scene graphs and employs these entities, along with some form of reasoning, to answer more complex types of queries about the visual input (Mao et al., 2019). NS-CL not only performs quite a bit better than the previous state of the art in the visual question-answering task but also provides interpretable reasoning chains as output.

Another approach along these lines is the "Logical Neural Network (LNN)" (Badreddine et al., 2022). This model directly incorporates first-order logic into the architecture of the neural network, allowing it to perform both arithmetic and logical deductions. The model remains an end-to-end trainable model for backpropagation. These approaches hold quite a bit of promise for applications where rules must be followed, such as regulatory compliance or legal automation. They have also been used to complete knowledge bases (somewhat of a "holy grail" for the AI community) (Besold et al., 2017).

They are somewhat similar to the architectures used in entity resolution but with less focus on pairwise relationships. As with algebraic structures, the constraints used in these architectures often embody reasoning similar to that of humans (Ji et al., 2021).

9.1.3 The Role of Symbolic Reasoning in Modern AI

Symbolic AI is grounded in logic-based systems, such as propositional logic, predicate logic, and ontologies. Such systems perform brilliantly in deductive reasoning, planning, and satisfying constraints. This makes them very useful in safety-critical areas, such as medical diagnostics. In such domains, known causal relationships (such as mappings from symptoms to diseases) are integrated into the systems that rule and drive them.

Autonomous Systems: Ensuring that robots or vehicles operate in accordance with logical rules and safety constraints.

Legal and Compliance Applications: In contexts where explainability and compliance with regulatory frameworks are essential, a key strength of symbolic reasoning is its support for commonsense reasoning. This enables a machine to use general knowledge and contextual information to draw inferences (Marcus, 2020). Table 9.1 summarizes the features of benefits and challenges in GenAI.

9.1.4 Research Outlook

Therefore, the next-generation AI will not stick to just one method but will blend brain-like nets with rule-based symbols, borrowing the strongest parts of both.

TABLE 9.1
Benefits and Challenges of Neural-Symbolic Integration in AI Systems

Aspect		Details
Benefits	Explainability	Symbolic modules offer human-readable reasoning traces.
	Efficiency	Reduces the volume of labeled data required through prior knowledge integration.
	Generalization	Enables zero-shot or few-shot reasoning in novel scenarios using logical rules.
Challenges	Integration Complexity	Harmonizing continuous (neural) and discrete (symbolic) representations is non-trivial.
	Computational Overhead	Reasoning modules can be computationally expensive, especially for large-scale symbolic spaces.
	Scalability	Symbolic methods may struggle with the vast and noisy datasets that deep learning handles well.

Although DL models are surprisingly good at recognizing patterns, they often miss simple things; therefore, they are impressive but not perfect. Thus, many call these models dumb; they hide their inner workings, and moreover, they never show any real thinking; how can we trust them? Unlike other models, many symbolic systems love logic and rule-based reasoning; therefore, they crumble whenever perception or adaptability occurs. Currently, researchers are trying to reduce this gap by stitching together hybrid models that learn from raw data; therefore, they still work with the tidy facts that humans use without thinking.

This shift was reshaped by fresh building ideas; therefore, it feels brand new. Logic is shoved right into the AI (a new trick); therefore, it learns all at once. **Rocktäschel and Riedel (2017)** built program-induction networks; thus, they learned algorithmic patterns from data and could evaluate symbolic expressions as well as generate synthesis code. Neural theorem provers combine DL tricks with logical rules to obtain step-by-step reasoning, make proofs, and even check them (Devlin et al., 2017). In addition, combining visual, language, and symbolic information in a hybrid model shows great promise for better generalization and clearer explainability. The 2018 results showed a sharp bump, so the numbers just jumped in that year (Rocktäschel & Riedel, 2017; Evans & Grefenstette, 2018). These systems can look at an image and match it to a short description; therefore, they think in a manner similar to humans when they choose a step and try to guess what could happen next. They could wonder which outcomes most humans would call right or wrong; therefore, they try to picture both. Imagine hybrid AI, not just copying data, so it aims at real thought: a system that gets what you ask, thinks it over, then does the right thing. Could these models increase trust? They may make things clearer and hold people accountable in healthcare, courts, and public policy. Reaching hybrid AI is a major checkpoint, and therefore, we are getting closer to AI that actually helps people. For example, hybrid AI systems can tutor children in mathematics or help keep neighborhoods safe (Marcus & Davis, 2019).

9.2 SCALABLE SOLUTIONS FOR FAIRNESS-AWARE SYSTEMS

Creating fair AI involves more than just bias-free outputs. In many cases, when we say that an AI system has fairness issues, we are letting it off the hook for producing truly problematic, even dangerous, outputs. From a fairness perspective, the outputs are only as good as the inputs and the model's ingredients. Therefore, this discussion is not just about making the model "do the right thing." It is also about giving the model "the right stuff" to work with. If you look up fairness in AI in the academic literature, you are most likely to find the definition put forth by Mehrabi et al. (2021). It is a good one. They state that fairness is multidimensional and often context-specific. An "unfair" output in one case might be "fair" in another case.

9.2.1 Techniques for Debiasing at Scale

There are three primary categories of debiasing methods as explained in Table 9.2.

Each method has trade-offs between performance, interpretability, and fairness outcomes. A comprehensive system often combines all three.

9.2.2 Key Challenges in Ensuring Fairness in GenAI

While numerous techniques exist, achieving fairness in large-scale generative models is challenging due to several systemic issues. Table 9.3 shows key challenges to ensure fairness in GenAI.

TABLE 9.2
Techniques for Debiasing at Scale

Debiasing Stage	Techniques	Example Methods
Pre-processing	Modify the input data before training to reduce bias.	Data augmentation, reweighting, sampling, feature balancing Kamiran and Calders (2012)
In-processing	Integrate fairness constraints or objectives during model training.	Adversarial debiasing, fairness-aware loss functions Zhang et al. (2018)
Post-processing	Adjust model outputs to ensure fairer outcomes.	Threshold adjustments, counterfactual fairness testing, calibration techniques

TABLE 9.3
Key Challenges in Ensuring Fairness in GenAI

Challenge	Description
Model Complexity	Models like GPT-4 or Stable Diffusion have billions of parameters, making it difficult to identify bias propagation paths.
Real-Time Bias Emergence	Generative models that interact with live user input (e.g., chatbots) may develop new biases that are not present in training data.
Tooling Gaps	There is a lack of standardized, scalable tools for auditing and testing generative model fairness.

TABLE 9.4
Tools and Approaches for Scalable Fairness Monitoring

Tool/Method	Functionality
SHAP (SHapley Additive Explanations)	Helps understand how input features affect a model's output, offering transparency.
Counterfactual Fairness Testing	Assesses if model predictions would change for different demographic inputs under identical conditions.
Fairness-aware Pipelines	Embedding fairness analysis and intervention directly within the generative model lifecycle.

9.2.3 Tools and Approaches for Scalable Fairness Monitoring

To scale fairness auditing and enforcement, several tools and strategies are being proposed. Table 9.4 shows tools and their functionality available for fairness monitoring.

The necessary tools should be multimodal and support various types of data, including text, images, and audio. They must also be real time, which means that they should keep constant tabs on not only the model but also the model's performance. While the ambiguity of "fairness" can lead to paralysis by analysis, it need not. With clearly stated and well-understood goals of what constitutes "fairness," the GenAI community can better address the complex issue of harmful model behavior and not just ignore it or hope it goes away. Indeed, with the community addressing harmful model behavior head-on, society at large can be more confident that GenAI tools are safe to use.

9.3 MULTIDISCIPLINARY FRAMEWORKS FOR ETHICAL AI

Creating ethical AI systems is not solely the province of the people who build AI. Their effect on nearly every part of human life means that building them requires an almost comprehensive and certainly collaborative understanding of many (or most) domains of knowledge. Audiences (human and otherwise) must be assured that AI will be created in such a way as to uphold not only human rights but also University of Pennsylvania (2017)-worthy cultural diversity and long-standing societal values. Building AI in this way requires a range of "societal stakeholders." Table 9.5 shows collaboration models for ethical AI.

Multidisciplinary collaboration is guaranteed.

- Equality and justice in global societies.
- Integrating the vast range of human diversity into all aspects of life.
- Compliance with laws and regulations and accountability for actions.
- Increased public confidence and open dealings.

TABLE 9.5
Models of Collaboration for Ethical AI

Model	Description	Examples
AI Ethics Board Model	Internal review boards within companies evaluate projects for ethical risk and compliance with values like fairness, privacy, and transparency.	Google's Advanced Technology External Advisory Council (ATEAC), Microsoft's AETHER Committee
Academic-Industry Partnerships	Universities and corporations work together to conduct research and implement guidelines that bridge innovation with ethical responsibility.	Partnership on AI (involving Apple, IBM, Facebook, etc.)
Citizen Panels/ Participatory Design	Engaging public communities and stakeholders in the AI design process to reflect user needs and lived experiences.	European Commission's citizen deliberations on AI
Public Ethical Frameworks	Broad charters or declarations created through multidisciplinary consultation to set shared ethical principles for AI development.	Montreal Declaration for Responsible AI (2018)

9.3.1 Case Example: The Montreal Declaration for Responsible AI

The Montreal Declaration for Responsible AI is one of the first attempts to establish ethical standards for AI and is certainly the most detailed attempt to date (Floridi et al., 2018). The reason for this is clear: Quebec and Canada have located themselves at the very center of the international debate concerning the future of AI, a position we hold because of the internationally recognized research carried out here on both the theoretical and applied frontiers of this domain. However, it is not only a detailed document; it is also highly participatory. The Montreal Declaration was not authored by policymakers working in a remote office (Gabriel, 2020). A public consultation held over several months and across many international events allowed a much larger and more diverse group of stakeholders to participate in the dialogue and record their opinions.

- *Respect for Autonomy*: It aids human decision-making. They do not allow for any reduction in human independence. These tenets govern the use of AI to assist humans in making decisions. They also serve as a base upon which the idea of responsible AI development can be built. These are noble, even aspirational, goals for any commissioner concerned about ethics. They are also very much related to the philosophy of AI. When we talk about using AI to help with human decisions, what kind of help are we talking about?
- *Protection of Privacy*: AI systems must comply with laws governing personal data. This means that secure handling of data is of utmost importance. AI systems must use data only with the clear consent of the individuals whose data are being used. AI systems must also ensure that the data are made

unidentifiable before being used to "train" the AI. The declaration in its "principles" section warns against surveillance. Practices that are shaping up to be surveillance in nature threaten a lot of what is good about using personal data in a way that benefits everyone.

- *Equity and Diversity*: This proclamation promotes justice and inclusion in every aspect of AI, from its conception to its impact on society. It makes no bones about the issue of bias—it says that bias or any appearance of bias must not be allowed to creep into these new technologies. What is probably my favorite part of the statement (and what makes it substantively different from a lot of the AI ethics Lip Service that gets paid) is that when talking about beneficial impacts, the statement's authors pay a lot of attention to how things play out for communities that are marginalized and historically oppressed. Indeed, the authors seem to care more about the oppressed than the oppressors of AI.
- *Democratic Participation*: Recognizing that AI can profoundly influence society, the declaration urges public involvement and open discourse on AI governance. It nudges states and organizations to involve the public in the decision-making process regarding AI. When they do, it pushes for those engagements to count for something and to involve as many citizens as possible in the conversation. Making ethics a part of AI design is not just the job of computer scientists. This is an interdisciplinary problem. If "creating a benevolent AI" is to become the default model for making AI research safe, then across different areas of expertise, that model should push for a kind of diversity that ensures a wide range of consequences (good, bad, or mixed) in AI's good–bad behavior.

The principles are as follows:

- Well-being
- Accountability
- Transparency
- Safety
- Responsibility
- Sustainable development

These tenets are established to align AI's growth with the public's best interests. It is important to emphasize that this is not just wishful thinking; this is also a very practical set of guidelines. They have already been applied to some real-life AI systems. The Montreal Declaration is worth looking into for several reasons. One is its content. What the Declaration actually says is quite significant. Another is "how" it was developed—if just "what" it says is significant, then how it got to be significant is worth examining.

Most of us would probably agree that there are much worse ways to arrive at a set of ethical AI guidelines than by collaborating with a community that thinks about this day in and day out. This is precisely how the Declaration on Responsible AI was formulated.

9.4 GOVERNANCE, POLICY, AND USER-CENTRIC DESIGN

The incorporation of AI into several critical areas of society is well underway and cannot be reversed. As of today, governmental and private authorities worldwide are busy trying to set up governance and policy frameworks for AI (Jobin et al., 2019). As AI is an emerging technology, the challenge for these frameworks is to strike a balance among four value-laden goals that we generally consider quite important: They need to promote (1) transparency, (2) accountability, (3) fairness—certainly a complicated, sometimes controversial, and delicate issue—and (4) favoring human rather than machine values.

9.4.1 Global and Regional Policy Frameworks

Numerous international initiatives have laid the groundwork for ensuring that the advancement of AI remains aligned with core democratic tenets and human rights issues. The first truly global regulation for AI, the EU AI Act, implements a risk-based framework that requires all AI systems, especially high-risk systems and those with potentially unacceptable risks (European Commission, 2021), to have human oversight and guarantees that they will be transparent and understandable. This is necessary because we are not far from a world that relies heavily on AI. Even now, we are seeing signs of an attempt to lay a technological foundation for what you might call a "world society," and it is not at all clear that these signs are leading us toward a society that is good, just, and free, or one that is not.

It identifies human rights as the most relevant, important, and crucial area affected by the development of AI. These rights are the basis for responsible AI. Almost every other kind of right and many obligations under international law and treaties are encompassed within the area of human rights. Thus, this framework has the widest possible scope.

9.4.2 National AI Strategies

Nations are swiftly adopting individualized strategies that correspond to their precise economic, cultural, and political priorities.

India—NITI Aayog's "AI for All" initiative aims to leverage AI for societal benefits in sectors such as agriculture, healthcare, and education. This program particularly emphasizes inclusivity and accessibility, and it is aligned with the skill enhancement activities of the National Skill Development Corporation to work toward eliminating the digital divide (NITI Aayog, 2018).

The United States—The White House Office of Science and Technology Policy published a framework for an AI Bill of Rights (Executive Office of the President, Office of Science and Technology Policy, 2022). It outlines five principles for protecting citizens that can serve as a starting point for discussion among various stakeholders regarding the types of ethical safeguards that should be put in place. Ensuring that AI systems are safe and effective before they are put into widespread use.

Ensuring algorithms do not discriminate: Why is this interconnected with general discriminatory practices, and why is it especially pertinent now? Discrimination can arise from several aspects in systems that use algorithms—sometimes in obvious ways and sometimes in ways that are not apparent. One of the clearest and most pressing conduits for potential algorithmic discrimination is the data fed into the systems. If these data are inherently biased, the algorithms are likely to reinforce and even amplify that bias, resulting in various adverse outcomes for the groups that are negatively impacted—outcomes that are very similar to those produced by a discriminatory process that does not use an algorithm (Raji et al., 2020).

Ensuring that AI systems provide clear and interpretable information about their processes and decisions enhances transparency, fosters user trust, and supports accountability in their deployment (Ding, 2018).

9.4.3 User-Centric Design Principles

Good AI governance requires not only effective policymaking at the highest levels but also an ethical approach centered on human needs, rights, and values that we ought to uphold.

- *Explainable Interfaces*: Outputs from AI systems must be understandable and trustworthy. If users cannot grasp the nature of the output, then the system fails them and their task (Doshi-Velez & Kim, 2017).
- *Consent Mechanisms*: In domains such as healthcare, finance, and consumer applications, we operate under "informed consent" frameworks, which allow humans to maintain control over their personal data (Mittelstadt et al., 2016).
- *Value-Sensitive Design*: With respect to what is arguably the most important VR issue (privacy), we must use "Value-Sensitive Design," achieving the aims of which makes for a well-designed stakeholder process (Friedman et al., 2006).

9.4.4 Technical Tools for Ethical Implementation

These governance ideals have prompted the development of several technical strategies and tools. The first is the Model Card, a standardized documentation tool that outlines a model's purpose, performance, and the kinds of ethical considerations that people should take into account when using the model. The second is Fairness Impact Assessments, which are used when AI is employed in high-stakes decision-making contexts, such as hiring or law enforcement. Then, there are our privacy-preserving techniques, which are the way to "handle" sensitive data without compromising the privacy of the individuals from whom the data were obtained. There is also multilayered governance that ensures that these first-layer technical tools do what they are supposed to do safely and fairly (Zliobaite, 2017).

9.5 FUTURE OUTLOOK AND OPEN CHALLENGES

The application of AI to many different kinds of problems has been buoyed by DL and generative models; however, several substantial and serious technical and philosophical problems have recently been highlighted, because either some critics have pointed them out or some responsible people within the AI community have seen the need to raise important issues. These substantial problems—of both sorts—need to be discussed and resolved. Otherwise, the picture being painted of the AI that is being created is one where we might end up with a "Frankenstein's monster."

- *Balancing Fairness and Performance*: Creating superior, equitable, and efficient AI systems is one of the most difficult things to accomplish because the overall performance of these systems can be reduced when they are made equitable. Despite this, and in light of recent events, it is important to emphasize that fairness cannot be neglected in the pursuit of performance. However, achieving performance and fairness can sometimes feel like one of the impossible tasks of Hercules, namely, cleaning the Augean stables. One unambiguously positive result of the effort to clean the stables is the realization that the performance of AI systems can sometimes be improved when they are made fair, a point not often acknowledged (Shneiderman, 2020).
- *Opacity of Black-Box Models*: Many advanced AI systems, particularly deep neural networks, are often described as "black boxes," owing to their lack of transparency. This makes it difficult for both users and developers of these systems to comprehend the bases of their decisions. While not understanding a system's decisions can make a user feel very uneasy, it can also have serious repercussions when a system is used to make decisions in high-stakes domains such as criminal justice and healthcare (Lipton, 2018).

  Although there are methods such as LRP and SHAP values that have been created to explain what a model is doing, obtaining full transparency in complex structures is still difficult.
- *Cultural Bias and Global Inequity*: Extensive datasets obtained from the internet or sourced from the dominant regions of North America and Europe are the basis on which AI models are built. However, because these domains of source material do not reflect a universally representative assembly of humanity, they are embedded into these datasets and the models they yield a certain cultural bias, inflected with the unequal social relations of the times we live in. And as Bender et al. (2021) state, "This practice may result in the systematic marginalization of minority groups, Indigenous communities, or languages that are not widely represented."
- *Long-term and Existential Risks*: Society faces serious long-term threats from the improper use of AI. This includes LLMs and deepfakes that can mislead a large number of people. They can also be used to influence public opinion and to hit societies with automated cyberattacks (Brundage et al., 2018). However, even when misused, generative models are not intrinsically poor. They are not harmful because they result in something harmful, such as ordering a drone to take out a target in your neighborhood. They are harmful because they cut at the most basic level of trust between the people and their government. When one cannot trust the signals one is receiving

from the government, it creates a level of unease that is bound to have serious repercussions for the society in which it is happening.

- *Future Directions: Toward Responsible AI*: Multidisciplinary research and governance frameworks must focus on value alignment, monitoring and auditing, and human–AI symbiosis to address concerns about AI. Value alignment is about ensuring that the AI systems we build are in accord with what we hold (society, along with ethical and legal authorities) as valuable; with what we are accustomed to using as the basis for setting standards; with what we reckon is legally binding; and with what the end users of the AI systems prefer. The realignment of values at the societal level, of authority, and of the basis for preferring one thing over another is the single most important thing. However, it is insufficient to have the correct values. We must also ensure that the systems function properly and do not cause harm. If they cause harm, we must determine the cause and prevent it. If we cannot do that, we will certainly not have AI systems that operate in accordance with our values.

9.6 CONCLUSION

There are two principal ways in which GenAI can advance. One way is to increase our technical sophistication. The other—equally important, in fact more important if we are to avoid potential dangers—is to ensure that it matures as a system of accountability, fairness, and trust in the system. Both are necessary for GenAI to be a successful tool. We now have a science of AI that is sufficiently good to model and that make decisions in complex systems with wide-ranging effects. However, despite its strengths, AI remains profoundly weak in terms of being understandable by humans. If we are to trust increasingly powerful AI systems, we must be able to comprehend them, at least in some basic, structural way.

The old hybrid strategy used by the best researchers—DL paired with a resurgence of symbolic reasoning—has never been more needed or more absent. Increased comprehensibility may also allow us to increase controllability.

EXERCISES

1. What are hybrid models, and how can they enhance the performance and interpretability of GenAI?
2. How can symbolic reasoning be combined with DL to improve the transparency and logic of generative systems?
3. What are the potential advantages of merging neural and symbolic approaches for developing responsible AI?
4. What current limitations need to be addressed to make hybrid models practically deployable?
5. Which research directions show the most promise for making GenAI robust and ethically aligned?
6. What strategies are being developed to scale fairness-aware solutions in GenAI?

7. How can fairness be maintained across models trained on dynamic or real-time data?
8. What challenges arise in scaling debiasing techniques for large generative models like LLMs?
9. How can organizations balance scalability, fairness, and computational costs?
10. What are the implications of deploying fairness-aware generative systems on a global scale?
11. Why is a multidisciplinary approach crucial for the development of ethical, GenAI systems?
12. How can collaboration among technologists, ethicists, legal experts, and social scientists improve AI governance?
13. What are some examples of successful multidisciplinary frameworks in ethical AI development?
14. How do cultural and regional differences impact ethical design in GenAI?
15. What role can partnerships between academia and industry play in shaping responsible AI ecosystems?
16. What governance mechanisms are essential to ensure the responsible development and use of GenAI?
17. How can user-centric design enhance trust, usability, and accountability in GenAI systems?
18. What role do national and international policies play in steering responsible AI innovations?
19. How can transparency and explainability be incorporated into AI governance?
20. What challenges and opportunities arise in establishing a globally harmonized AI policy framework?

REFERENCES

Badreddine, S., d'Avila Garcez, A. S., Besold, T. R., & Lamb, L. C. (2022). Logic tensor networks. *Artificial Intelligence, 301*, 103574.

Barocas, S., Hardt, M., & Narayanan, A. (2019). *Fairness and machine learning: Limitations and opportunities.* fairmlbook.org.

Bender, E. M., Gebru, T., McMillan-Major, A., & Shmitchell, S. (2021). On the dangers of stochastic parrots: Can language models be too big? In *Proceedings of the 2021 ACM Conference on Fairness, Accountability, and Transparency* (pp. 610–623).

Besold, T. R., et al. (2017). Neural-symbolic learning and reasoning: A survey and interpretation. *Frontiers in Artificial Intelligence and Applications, 285*, 1–10. *(Note: Duplicate arXiv version removed.)*

Brundage, M., Avin, S., Clark, J., et al. (2018). The malicious use of artificial intelligence: Forecasting, prevention, and mitigation. *arXiv preprint arXiv:1802.07228.*

Devlin, J., Uesato, J., Bhupatiraju, S., Singh, R., Mohamed, A., & Kohli, P. (2017). RobustFill: Neural program learning under noisy I/O. In *Proceedings of the 34th International Conference on Machine Learning* (pp. 990–998).

Ding, J. (2018). *Deciphering China's AI dream.* Future of Humanity Institute, University of Oxford.

Doshi-Velez, F., & Kim, B. (2017). Towards a rigorous science of interpretable machine learning. *arXiv preprint arXiv:1702.08608.*

European Commission. (2021). *Proposal for a regulation of the European Parliament and of the Council laying down harmonised rules on artificial intelligence (Artificial Intelligence Act) and amending certain Union legislative acts* (COM(2021) 206 final). Brussels, Belgium: European Commission. https://eur-lex.europa.eu/legal-content/EN/TXT/?uri=CELEX:52021PC0206

Evans, R., & Grefenstette, E. (2018). Learning explanatory rules from noisy data. *Journal of Artificial Intelligence Research, 61*, 1–64.

Executive Office of the President, Office of Science and Technology Policy (OSTP). (2022). *Blueprint for an AI bill of rights: Making automated systems work for the American people.* The White House. https://www.whitehouse.gov/ostp/ai-bill-of-rights/

Floridi, L., et al. (2018). AI4People: An ethical framework for a good AI society. *Minds and Machines, 28*(4), 689–707.

Friedman, B., Kahn, P. H., Jr., & Borning, A. (2006). Value sensitive design and information systems. In P. Zhang & D. Galletta (Eds.), *Human-computer interaction and management information systems: Foundations* (pp. 348–372). M.E. Sharpe, Inc.

Gabriel, I. (2020). Artificial intelligence, values and alignment. *Minds and Machines, 30*(3), 411–437.

Ji, S., Pan, S., Cambria, E., Marttinen, P., & Yu, P. S. (2021). A survey on knowledge graphs: Representation, acquisition, and applications. *IEEE Transactions on Neural Networks and Learning Systems, 33*(2), 494–514. https://doi.org/10.1109/TNNLS.2021.3070843

Jobin, A., Ienca, M., & Vayena, E. (2019). The global landscape of AI ethics guidelines. *Nature Machine Intelligence, 1*(9), 389–399.

Kamiran, F., & Calders, T. (2012). Data preprocessing techniques for classification without discrimination. *Knowledge and Information Systems, 33*(1), 1–33.

Lipton, Z. C. (2018). The mythos of model interpretability. *Communications of the ACM, 61*(10), 36–43.

Mao, J., Gan, C., Kohli, P., Tenenbaum, J. B., & Wu, J. (2019). The neuro-symbolic concept learner: Interpreting scenes, words, and sentences from natural supervision. In *Proceedings of the International Conference on Learning Representations (ICLR 2019).* OpenReview.net.

Marcus, G. (2020). The next decade in AI: Four steps towards robust artificial intelligence. *arXiv preprint arXiv:2002.06177.*

Marcus, G., & Davis, E. (2019). *Rebooting AI: Building artificial intelligence we can trust.* Pantheon.

Mehrabi, N., Morstatter, F., Saxena, N., Lerman, K., & Galstyan, A. (2021). A survey on bias and fairness in machine learning. *ACM Computing Surveys (CSUR), 54*(6), 1–35.

Mittelstadt, B. D., Allo, P., Taddeo, M., Wachter, S., & Floridi, L. (2016). The ethics of algorithms: Mapping the debate. *Big Data & Society, 3*(2), 2053951716679679.

NITI Aayog. (2018). *National strategy for artificial intelligence #AIForAll.* Government of India. Link.

Raji, I. D., Smart, A., White, R. N., Mitchell, M., Gebru, T., Hutchinson, B., . . . & Barnes, P. (2020). Closing the AI accountability gap: Defining an end-to-end framework for internal algorithmic auditing. In *Proceedings of the 2020 Conference on Fairness, Accountability, and Transparency* (pp. 33–44).

Rocktäschel, T., & Riedel, S. (2017). End-to-end differentiable proving. In *Advances in neural information processing systems* (pp. 3788–3800). Neural Information Processing Systems Foundation.

Shneiderman, B. (2020). Human-centered artificial intelligence: Reliable, safe & trustworthy. *International Journal of Human-Computer Interaction, 36*(6), 495–504.

University of Pennsylvania. (2017). *AI + society: Cultural and societal considerations.* University of Pennsylvania. https://ai.upenn.edu/ai-society/

Yi, K., Wu, J., Gan, C., Torralba, A., Kohli, P., & Tenenbaum, J. (2018). Neural-symbolic VQA: Disentangling reasoning from vision and language understanding. In *Proceedings of the 32nd International Conference on Neural Information Processing Systems* (pp. 1031–1042).

Zhang, B. H., Lemoine, B., & Mitchell, M. (2018). Mitigating unwanted biases with adversarial learning. In *Proceedings of the 2018 AAAI/ACM Conference on AI, Ethics, and Society* (pp. 335–340).

Zliobaite, I. (2017). Measuring discrimination in algorithmic decision making. *Data Mining and Knowledge Discovery, 31*(4), 1060–1089.

Index

Note: Page locators in **bold** indicate a table and page locators in *italics* indicate a figure.

academic integrity 28, 38, 50, 164
accountability: concerns 3, 22–23, 40–41; ethics 113–115, 117; gap 35, **36**; lack of 32, 130; legal 50; technological 28, 109
accuracy: diagnostic 75, **154**; high **62**, **72**, 141, 146, 148; model 14; overall 135, 146; trade-offs 107, 111, 118, 144, 146
activation: functions 14–15, **15**, 16; maximization 119, **122**
active learning 73–74
Advantage Actor-Critic (A3C) 67–68
adversarial: debiasing 96, 98, **129**, 135; learning 5; loss **65**; neural networks 83 (*see also* generative adversarial networks); training **14**
adversarial debiasing 96, 98, **101**, **129**, 135, 164
African American Vernacular English (AAVE) 96–97, **98**
AI Bill of Rights 168
AIF360 107
algorithmic: auditing 156; bias 37, **98**, 141, 149; debiasing 3, 98, **101**, 110, 132; design 1, 110; discrimination 95, 110, 169; learning 1, 70
alignment 7, 22–25, 107, 135, 161, 171
AlphaFold Protein Structure Database (AlphaFold DB) 29
Amazon 149
anti-discrimination laws 149, 153
ARIMA 61
art: creation (image) **5**, 28, 30, 44–45, **87**, 89; of explanation 118; forms **48**
artificial intelligence (AI): applications 124, 128, 140, 151; biased 132, 135; concepts 8; evolution of 1, **2**, 4, *6*, 59; fair-explainable system 141; functions of 7; good-bad behavior 167; high-risk 124, 132, 168; learning paradigms *60*; pillars 8, 23–24, 60; supervised learning *60*
Artificial Intelligence Act 148
artificial neural network (ANN) 14–16
artistic creativity 45, 100, 119
artists 30, 34, 37, 44–46, 50
attention mechanism 19–21, 23–24, 121, **122**
audio data, generation models 1, 3, 7, 9, 22, 63, **69**, 70, 74, 84, **87**, 113, 165
audits **98**, 122, 140, 155–156; *see also* fairness
authorship 33, **36**, 45, 50, 52
autoencoder 16–18, 61, 84–85, **88**, 91
automatic speech recognition (ASR) 18, 63, **66**, 70
automation 3, 27, 30, 37, 43, **87**, 91, 150–151, 162
autonomous: systems 90, 150, 162; vehicles **67**, 76
autonomy 52, 108, 151, 166

backpropagation 15
balance 28, 35, 42, 48, 67, 95, 119, 139, 146
Banerjee, I. (et al 2021) 103, **104**
banking 27, 49, **138**; bias in lending 33, **36**, 49, 138, **140**
barriers 31, 42, 45, 118, 146, 148, 150, 152, 156
benchmarks 105, 124, 141, 153–154
Bender, E.M. (et al 2021) 36, 101, **104**, 170
beneficence 109
BERT (Bidirectional Encoder Representation from Transformers) 69, **72**, 78, 86, **88**, 147
bias: amplification 96; case studies 101, **104**; in data descriptions 96; diagnosis 136; hidden 39, 48, 124, 129–130, 133; historical 94, **98**; intersectional 95, 97, **98**; linguistic/cross cultural 97; and mitigation process *130*; systemic 97
bilingual evaluation understudy (BLEU) 21, 24
biomedicine 49
Birhane, A. (et al 2022) 101, **104**
black-box: fairness 106; opening 131, 136; problem 34
Bryson, A. 102

Caliskan, A. 102, **104**
cancer 37, 139
Chatbots **5**, 6–7, 38–39, 40, 42, 81, **87**, 90, 102, **104**
ChatGPT **5**, 6, 11, 30, 50, 70–71, 93, 97
Chesney, R. **36**
choices 27, 38, 44, 50–51, 76, 108, 124, 128, 151–152
Christie's 30
Citron, D.K. **36**, 40
clarity 49, 115, 123, 129–130, 133, 152
class distribution 94–95
Claude **5**, 7, 30
CLIP 5, **5**, 44
clustering 8, 61, **64**, **72**, 75, 77
communication 41–43, 103, 118, 147, 161
comparative synthesis 48
computed tomography (CT) scans 37
computer: code 30; science 7, 27–28; vision 7, 69–71, 86, 92
consent 34, **36**, 43, 49, 51, 91, 100, 149, 166, 169

constraints 17, 122, **138**, 147, 162, **164**
context: creative 13, 91; cultural 152; human 22, 43, 71, 133–134; legal 110, 150, 162; privacy 48, 160; standards 155
conversational bots *see* chatbots
convolutional neural network (CNN) 1, 15–16, 60, 75, 84–85, 162
copyright: issues 3, 90; violations 33
CORD-19 32
counterfactual: explanation 122, 131; fairness **101**, 105–106, **164–165**
COVID-19 30–31
creative: content 4, 90–91, 117; control 27, 30; economy 34, **36**, 45; industries 34, 44–45, 50, 52
credit scoring 38, 60, 115, 132, 150
criminal justice 113, 136, 141, 150, 170
crisis management 32, 42
critical thinking 27, 31, 38, 50, 116
criticisms 123–124
cross-entropy loss 14–15, 21, 23–24
cultural: bias 94, 97, **98**, 170; homogenization 46, 50; shift 93, 110, 141, 150–152, **154**
customer service 90

DALL-E **2**, 3, **5**, 5–6, 9, 11, 30, 87, **87–88**, 93, 97, 101, 117
data: augmentation 24, 69, **72**, 89, 91, 95, **164**; collection 27, 133; distribution 21, 61, **65**, 70, 85, 87; high-dimensional 61, **62**, **66–67**, 67, 126, 161; integrity 107; processing layer 10, 82; protections 34, **36**, 108, 115, 149; synthesis 71; synthetic 34, **36**, 37, 49, **71**, **84**, **90**
datasets: diverse 10, **36**, 95, **140**; labeled/unlabeled 63, 68, **71–73**, 73, 75; large 1, **2**, 3, 9, 11, 22, 25, 32, **64**, **66**, **68**; training 22–23, 136, **140**, 151
debiasing at scale 164; *see also* adversarial; algorithmic
debugging 31, 115
decision trees **2**, 60, **62**, **72**, 77, 118, 120, 122, 124, 145
decoder 17, 19–21, **64**, 85, **88**
Deep Deterministic Policy Gradient (DDPG) 67–38
deep learning (DL): multilayered function **2**, 3, **5**, 9; networks 128–129; research 124, **163**
deep neural network (DNN) 4, 11, 15, 67, 117
Deep Q Networks (DQN) 67, **72**, 76
deepfakes 22, 24, 32–33, **36**, 43, **48**, 49–50, **88**, 170
demographic parity 99, **101**, 103, 105, 110, **138**, 153
denoising autoencoder (DAE) 18, **88**
design principles 156, 169
diagnostic tools 103
dialogue 31, 89, 166
diffusion models **14**, 27, 71, **72**, 87–88, **88**, 91
digital divide 50, 151, 168
Digital Personal Data Protection (DPDP) 108
discrimination: algorithm 95, 110, 132, 135, 149, 168–169; credit bias 33, **36**, 38; social 4, 27
Discriminator (D) 83–84, **88**, 91
disease 29, 60–61, **62**, 74–74
divergence *see* Kullback-Leibler
diversity 21–22, 31, 37, **65**, 93, 153, 165, 167
DNN deep neural network
drug discovery 3, 29, 37, 47, **48**, 86, 89

education: pros and cons 38, 50; student clustering 75; writing assistants 102
efficiency 19, 24, 30, 37, 40, 48, 61, 71, 73, 148, **163**
embedding 19–20, 27, 49, **64**, 95, 106–107, **165**, 170
encoder 17–21, **64**, 69, 85, **88**
engineer 3, 94, 102, **104**, 113, 133, 154
engineering **2**, 28, 30, 81, 118
entertainment 8, 22, 89
environmental: impact 23, 32, 35; sustainability 147
equalized odds 105, 111
equitable access 98, 110, **129**, 135, 151, 170
equity 3, 38, 48, **101**, 155, 167, 170
ethical: challenges 24, 32, **36**; concerns 27, 31–32, 40, 50, 130, 151; consideration 22, 52, 114, **154**, 169; implications 49, 52, 169; landscape 27–28; negligence 109; principles 52, 107–108, 155, **166**; trade-offs 35, **36**
ethicist 28, 94, 172
EU AI Act 35, **36**, 107, 132, 141, 168
European Central Bank (ECB) 49
European Union (EU): AI guidelines 28; legislation 107, 115 (*see also* EU AI Act)
evaluation: cross-domain 154; gaps 152
evaluation metrics 11–12, 70, 141, 153
explainability: in audits 130; example-based 123–124; flowchart 41, *125*; in GenAI 3, 113, *116*, 119, 124, 141; intrinsic approach 114, 122, 124; methods **122**, 133; obligations 132; research 124; synthesis/outlook 124
explainable AI (XAI) 35, **36**, 116–117, 123, 125, 134–135, 140–141
explainable GANs (XGANs) 121, **122**, 126
explanations: counterfactual 122, 131; narrative-based XAI 123; natural language 123–124

fairness: approaches and techniques 3, 93, 99–100, **101**; audits **101**, **104**, **129**, 130, 137, **138**, 139, **140**, 141–142, 165; concepts of 93, 110, **129**, 146; and explainability 3–44, 11, 128, **129**, 134–135, 141–142; legal framework 107, 150; role of 24, 94; scalable 15, 164–165; standardization 153, 155–156; trade-offs 107, 111, 144, 153, 164
fairness-aware systems **36**, **98**, 136, 164, **165**, 171–172

features: image 12, 69, 75; input 14, 121, **122**, **165**; key/significant **2**, 17, 84
federated learning 73, 75
feed-forward neural network (FFN) 19–20
FID *see* Frechet Inception Distance
finance: applications 73, 89–*90*; domain specific 118; explainability **138**; pros and cons 37–38, **48**, 108; risk 4, 49, 146, 151; transparency 114, 150
fine-tuning 68, **71**, 73, 81, **98**
foundation model 10–12, 14, 24, 106
fraud detection 37, **48**, 49, 60–61, **62**, 90
Frechet Inception Distance (FID) 22, 24–25, 70
fuzzy 51, 93, 124

Ganguli (et al 2022) 102, **104**
gated recurrent unit (GRU) 19
GDPR *see* General Data Protection Regulation
gender: bias 101–103, 105, 130–132, 135, 137, **138**, 144; fairness 105–106, 106; gap 33; representation 95–96; stereotypes **36**, **104**
General Data Protection Regulation (GDPR) 34, **36**, 108, 115, 125, 132, 149–150
Generative Adversarial Networks (GANs): algorithms **65**, 95; bias **98**; development 5, 81; ethics 27; explainability 113, 117, **122**, 124; latent space exploration 120–121; learning paradigms 70–71, **72**, **81**; probability 13, **14**; transformer-based 86, 91
Generative artificial intelligence (GenAI): advantages 42, 44, 46; assistants 30; bias types **98**, *100*; can act on it's own 27; challenges 43–45, 47; comparison table **88**; creative advantages 43; data transformation 9, *10*; ethics 22, 49; evaluation metrics 21; explainability 113–114, *116*, 117, 119, *125*; five-layered architecture 81–82; functions 7, 27; information overload 44; moral use 23, 27; promise of 28, *29*; pros and cons **48**; real world applications 89
generative learning 59, 70–71, 78, 110
generative model: bias 22, 52; challenges 164; data distribution 61; evaluating 21; foundations 4; image generation 12; layer 81; medical 103; misinformation 22
Generative Pre-trained Transformers 5
generative reinforcement learning (GRL) 71, **71**
GitHub Copilot **5**, 6, 31
global: challenges 31, 35; data protections laws 108; explainability scores (XGMs) 124; regulation 35, 168
Global North 151
Gmail 102, **104**
Goodfellow, I. 5, 27
Google 5, **5**, 9, **87**, 102, **104**, 161, **166**
governance: frameworks 23, 28, 110, 171; good 36; model 109, 142; policy stimulation 39
government/law: advantages of AI 38–39; challenges 40; citizen engagement 40; policymaking 39; surveillance 41, **48**
GPT (Generative Pre trained Transformer): GPT-3 5–6, **87**, 101, **104**, 118, 146; GPT-4 7, 30, 33–34, **36**, **87**, **164**; models **2**, 3, **5** (*see also* OpenAI)
graphics processing unit (GPU) 1, **2**, 23–24, 146–147

Han, S. (et al 2015) 148
Health and Human Services (HHS) 114
Health Insurance Portability and Accountability Act (HIPAA) 34, **36**
healthcare: diagnostics **48**, 139; disease diagnosis 74–75; imaging 37, 49, 63, 75; labeled and unlabeled scans (MRI) 75; medical reports 76; patient privacy 37, 49 (*see also* HIPAA); pros and cons 37; records 34, 49, 141
hiring algorithm 137, **138**
HITL *see* human in-the-loop
human: brain 4, 14, 37, 47, 161–162; cognitive limitations 118; evaluation 21–22, 24; intelligence 1, 7, 116; judgment 22, 40, 116, 134; oversight 40, 94, 132, 140, 142, 168; rights 28, 35, **36**, 43, 50, 108, 115, 165, 168; values 22, 28, 91, 122, 151
human in-the-loop (HITL) **129**, 133–134, 142
hybrid: AI 161, 163; learning 73; models **65**, 73, 74, 124, 146, 156, 161

IBM: AI Fairness 360 100, **101**; AIF360 107; Deep Blue 4, **5**; Neuro-Symbolic Concept Learner (NS-CL) 162
image: compression 17, 61; features 12, 69, 75; generation 3, **5**, 13, 24, **71**, 86, **87–88**, 91, 101, **104**, 106, 120; medical 60, **62**, 63, **66**, **71**, 75–76, 89, 134, **140**; pixels 136; quality 22, 24; recognition 9, 15, 59–60; representation **69**, 76; restoration 18; synthesis 15, 70, 81, 86, **87**, 89, 123; unlabeled 75–76
inception score (IS) 22, 70
inclusion 3, 38, 49, 93, 167
Independent Component analysis (ICA) 61
India 108, 168
individual fairness 106
inequality 28, 34, **36**, 40, 102, 151, 156
inequitable access 151
innovation 7, 20, 22–23, 31, **36**, 45, 86, **101**, 115, 150, **166**
input training 10
integration 82, **101**, 121, 142, **163**
intellectual property (IP) 27–28, 33, 35, **35–36**, 45–46, **48**, 52, 149
interfaces 82, **140**, 169
interpretability 23, 71, 74, 77, **88**, 118–119, 122, 131, 139, **154**, 156, 165

Japan 152
joint distribution 13, **14**
journalism 31, 33, **36**, 43–44, **48**, 50, 116
judicial AI bias 40
justice 3, 36, 43, 52, 93, 108, 113; *see also* criminal

KL divergence *see* Kullback-Leibler
Kullback-Leibler (KL) divergence 21, 23–25, **65**, 85–86, 91

label 15, **66**, **69**, **72**77, 84
language: barriers 42; gendered 102; models/modeling 3, **5**, 8, 21, 30, 69–70, **104**, 120 (*see also* large language model); natural 6–7, 92, 123–124 (*see also* natural language processing)
large language models (LLM): advanced AI 9 (*see also* black box); bias 33, **36**; development 5, 8, 81, **87**; functions of 89–90; personal information leaks 34, **36**
Lasso 61, **62**
latent space: effective 11, 85; explicit/implicit **88**; exploration 84, 120, **122**; fairness 106; interpolating 120; lower-dimensional 85
Layer-wise Relevance Propagation (LRP) 120–121, **122**, **126**, **170**
layers: independent vertical stack 82–83; maintenance/monitoring 82
Le, Q. 148
learning algorithms: comparison charts **72**; continual/lifelong learning 73–74; contrastive learning **69**, **72**, 76; self-supervised 76
Learning Management Systems (LMS) 75
legal: agreements 30; assistance 39, **48**; barriers 148; consequences 109; context 110, 150; framework **36**, 107, 115, 150; obligations 132
legitimacy 40
linear regression 60–61, **62**, **72**, 118, 120, 145
loan approval algorithm **62**, **138**
Local Interpretable Model-Agnostic Explanations (LIME) 100, **101**, 107, 116, 119–120, **122**, 124–126, **129**, 131, 134, 136, **138**, 139, 146, 153
Logical Neural Network (LNN) 162
long short-term memory (LSTM) 18–19, 60–61
loss function 12, 14–15, 18, 21, 23–24, 85, 96, **98**, **164**

machine learning (ML): algorithms (supervised) **62**; density-based methods 61, **64**; statistical 161
magnetic resonance imaging (MRI) 37, 63, 75, 89
manipulation 24–25, 43, **48**, 50
marketing 30, 61, **87**, 91, 93
masked language modeling 69–70
mean squared error (MSE) 14–15, 23
media: advantages of AI 42; challenges and risks 43; misinformation 43 (*see also* deepfakes)
Mehrabi, N. (et al 2021) 164
memory 18–19, 30, 146
meta-learning (learning to learn) 74
metrics 11, 12, 21–22, 70, 98, 103, 106, **129**, 153–154
Midjourney **5**, 6, 87, 93
Miller, T. 118
misinformation 22, 24, 27, 31–33, **36**, 43, 51
misuse 28
mitigation 82, 95, *96*, **98**, 100, 110, 130, 134, 148
mitigation strategy 95, *98*, 148
MoCO (Momentum Contrast) 69
model: operational layer 82; training 18, 24, 73, 81, **101**, 102, 110, **164**; transformer-based 3, **5**, 34, 69, 86, **87**, 91–92, 126
model-agnostic 120, **122**, 124–125, 153
molecules 29, 32, 46–47, 51, 89
Montreal Declaration for Responsible AI 166
moral obligation 132
MRI *see* magnetic resonance imaging
multimodal learning **2**, 73–74
Multiple Choice Questions (MCQs) 23, 25, 77
music and sound design 89
musicians 30, 45
MYCIN 1, **2**

Naïve Bayes 60, **62**
Narayanan, A. 102–103
National Skill Development Corporation 168
natural language processing (NLP) 1, 7, 19, 21
Neuman, A. 133
neural network: architecture 14, 69; concepts 4, **5**, 15, 74; deep (*see* DNN); models 4, **5**, 11; multilayer **2**, 145; trained 10, 15
Neuro-AI 121
neuro-symbolic: integration **163**; models 74
Neuro-Symbolic Concept Learner (NS-CL) 162
New York City Local Law 144 108
non-compliance 109
non-fungible tokens (NFTs) 45
non-maleficence 109

Obermeyer, Z. (et al 2019) 103, **104**
object detection 15, 60, 68–69, 86
opacity 121, 128–129, 141, 151
opaque 119, 141, 150, 154
Open AI 5–6, 9, 30, 68, 102, **104**
Organisation for Economic Co-operation and Development (OECD) 28
output: auditing 23; generator 13, 84; harmful 35, **36**; novel 9, 11; production of 3, 15
oversight 28, 31–32, 47, 50, 52, 107, 133, 152; *see also* human
oversimplification 119

Partial Differential Equation (legal document) 115
pattern recognition 59, 74
Perceptual Path Length (PPL) 70
performance: metrics 81, 139; model 61, 63, 82, 110, 126
perplexity 21, 23–24
Personal Information Protection and Electronic Documents Act (PIPEDA) 108
physics 30, 46
policy: frameworks 168, 172; simulation 39
policymakers 4, 23, 28, 39, 114, 156, 166
prediction: accurate 59, 105, 136; final 14, 21; model 82, **138**, 139, 141, 145, **165**
predictive power 118, 141
principle component analysis (PCA) 17, 61, **64**, **72**, 77
privacy: data 23, 48, 51, 73; preservation 90; protection 37, 73; rules 34, 51; surveillance concerns 41
probability distribution 12, 14, 21, 23, 70, 85–86
productivity 7, 28, 30–31, 34
prototypes 23–124, 121, **122**, 123
Proximal Policy Optimization (PPO) 67–68
public: acceptance 133, 151; opinion 43, 152, 170; trust 27, 40, 50, 151; values 23

Q-learning algorithm 67

racial bias 95, 101–102, **104**, 139, 144, 149
random forest 60–61, **62**, 124
re-sampling 136
re-weighting **129**, 136
reasoning: logical 74, 121, 162; symbolic 122, 124, 161–162, 171
recall-oriented understudy for gisting evaluation (ROUGE) 21, 24
rectified linear unit (ReLU) **15**, 16
recurrent neural network (RNN) 1, 18, 86
Reddit 102, **104**
regression 8, 14; linear 60–61, 118, 120, 145
regulations 23, 28, 41, 43, 46, 107–108, 165
regulatory: compliance 90, 115, 140, 162; frameworks 28, 141, 162
reinforcement learning (RL) 63, **67**, 67–68, 70–71, 78, 82, **104**
reinforcement learning from human feedback (RLHF) 23, 71, **104**
representation parity 106
research: opportunities 161; scientific 28–29, 46–47
Riedel, S. 163
risk: education 38; elimination 36; models 49, 90; non-compliance 109; of oversimplification 119; of technology 40–41, 44–45, 47, 49
Rocktäschel, T. 163

saliency maps 117, 120, **122**, **129**, 136, 139, **140**
Samuel, A. 1
Samuelson, P. **36**, 45
scalability 1, **2**, **66–67**, 82, 117, 146, **163**, 172
Scaled Dot-Product Attention 20
Schwartz, A. (et al 2020) 148
Science, Technology, Engineering, and Mathematics (STEM) 148
scientific: discovery 28–29, 91; research 28–29, 46–47
scriptwriting 89
self-attention mechanism **2**, 5, 19, 86
self-supervised generative learning (SSGL) 71
self-supervised learning **64**, 68, **69**, 70–71, 78
semi-supervised learning (SSL) 63, **66**, **72**, 75, 77
sentiment analysis 60, **66**, 69, 87
Seyyed-Kalantari, L. (et al 2021) 103, **104**
SHAP *see* Shapley Additive Explanations
Shapley Additive Explanations (SHAP) 100, **101**, 107, 115, 119–120, **122**, **129**, 131, 136–137, **138**, 146, **165**, 170
SimCLR (Simple Contrastive Learning Representation) 69, **72**, 76
social: acceptance 151; barriers 150, 152; media 33, 42, 63, 107
software development 7–8, 31–34, 42, 45, 51, 84, 128, 152
South Korea 152
spam email 60 **62**, 128
Stable Diffusion **2**, **5**, 6, 30, 33, 87, **88**, 91, 118, 126, **164**
standardization 152–153, 155–153
storytelling **87**, 106
supercomputer 154
support vector machines (SVM) **2**, 60, **62**, **72**, 145
surge 27, 32, 43
sustainability 29, 31, 82, 146–147
Symbolic AI 1, **2**, **5**, 74, 121, **122**, 124, 161–162
symbolic reasoning 122, 124, 161–162, 171
synthetic data 34, 37, 49, **71**, 84, 90

Tan, M. 148
tasks: data pattern recognition 59, 74; generative 91, 110; labor related 34; legal 39; prioritize 19; text generation 21; text/pretext; vision 86
technology: beneficial 86, 100, 114, 119, 130, 153, 167; development 23, 152
text: generation 5, 18, 21, 24–25, 70, 86, **87–88**, 101, 121–122; summarization 69
trade-offs: ethical 35, **36**; fair/accurate 144, 146, 153, 164; hidden 39
training: bias 33, 94, **101**; datasets 22–23, 136, **140**, 151; fairness-aware **36**, **98**, 136; foundation model 10, 18, 24, 81, **101**, 110, **164**
transfer learning **72**, 73, 78
transformer model: applications 86–87; architecture *20*; components 19–20; core innovation 20

transparency: crucial 90; financial 49; lack of 90, 125, 129, 170
trust: issue of 3, 150, 152, 170; public 27, 40, 50, 150–151
Twitter (X) 102

United Nations Educational, Scientific, and Cultural Organization (UNESCO) 28
United States White House 168
University of Pennsylvania 165
unsupervised learning 61, 61, 63, **64**, 70, **72**, 73, 75, 78
uptake 116
US Algorithmic Accountability Act 108
US Federal Reserve 49
user: acceptance 116; bias 97, **98**, 102; data 90; end user 125, 130, 152, 171; engagement 81, 152
user-centric design 168–169

value-sensitive design 169
variational autoencoders (VAEs): applications 84–85, **88**, 91; block diagram *85*; divergence (KL) 21, **65**; learning algorithms 71, **72**; models 3, 10, 13, **14**, 113–114, 120, 124; probability distribution 70
Vaswani (et al 2017) 19, 27
vector 17–18, 20, 28, 60, 84, 120, 145
video: creation 7, **87**, 90; editing 42, **87**, 90; generation **5**, 74, 84, **87**
virtual: assistant 8, 31, 38, 40, 90, 106; characters 89
Vision Transformer (ViT) 76, 86, **87**
visualization 4, 77, **87**, 117, 119, **122**

Word Embedding Association Test (WEAT) 106–107
workflows 42, 78, 89
workforce 96, 147, 151
writers 30, 45, 102

XGAN 121
XGBoost 61, **62**

YOLO (You Only Look Once) 60

For Product Safety Concerns and Information please contact our EU
representative GPSR@taylorandfrancis.com
Taylor & Francis Verlag GmbH, Kaufingerstraße 24, 80331 München, Germany

www.ingramcontent.com/pod-product-compliance
Ingram Content Group UK Ltd.
Pitfield, Milton Keynes, MK11 3LW, UK
UKHW020204200926
475702UK00004B/601
* 9 7 8 1 0 4 1 1 9 7 3 5 5 *